Recent Climatic Change

Recent Climatic Change
A Regional Approach

Edited by S. Gregory

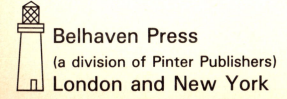

Belhaven Press
(a division of Pinter Publishers)
London and New York

© Stanley Gregory, 1988

First published in Great Britain in 1988 by
Belhaven Press (a division of Pinter Publishers),
25 Floral Street, London WC2E 9DS

British Library Cataloguing in Publication Data

A CIP catalogue record for this book is available from the
British Library

ISBN 1 85293 010 1

Library of Congress Cataloging in Publication Data

Recent climatic change.

    Bibliography: p.
    Includes index.
    1. Climatic changes.   I. Gregory, S. (Stanley)
PC981.8.C5R4 1988   551.6   88–10533
ISBN 1–85293–010–1

Filmset by Mayhew Typesetting, Bristol, England
Printed by Biddles of Guildford

Contents

List of tables

List of figures

Introduction

Studies of recent climatic change -- setting the scene

S. Gregory University of Sheffield, United Kingdom

The scientific acceptance that climates in the past were different from those of the present, and that therefore climate had been a changing phenomenon, came first in terms of the geological time-scale, as the existence of previous ice ages was acknowledged during the second half of the nineteenth century. The realisation that significant changes had occurred within the historical period itself was delayed in large measure until the first two decades of the present century. Then the writings of such as Kropotkin (1904), Bowman (1909), Huntington (1907; 1912; 1914; 1915; 1916) and Gregory (1914) explored both the evidence for such changes and their implications for societies in the past. It it true, however, that the rare enquiring spirit had raised these questions almost a century earlier (e.g., Schouw 1828).

Such considerations and discussions continued into the 1920s and 1930s (e.g., Brooks 1928; 1931; Bowman 1935), but by this time the statistical analysis of reliable instrumental data was demonstrating marked changes in climatic elements over periods of several decades or less, quite apart from the obvious fluctuations from one year to the next. The contemporary journals display this development most effectively. By mid-century it was possible to draw on a very wide research literature to establish changes both during the previous few centuries in general (e.g., Manley 1950; 1953; 1959; 1974) and also within the instrumental period itself (e.g., Ahlmann 1948; 1953; Manley 1944; 1949; 1951).

Over the past 30–40 years there has been a veritable explosion of research and publication in this field, culminating in the intensive activity of the past decade. Part of this has continued to be concerned with the potential impact of changing climate upon human history, as reflected in the volumes of Ladurie (1971), Lamb (1977; 1982) and Wigley *et al.* (1981). In turn, the concern with the possible impact of future climatic changes upon society and the economy is seen in such as Kates *et al.* (1985), which illustrates the growing role of international bodies in this field — the Scientific Committee on Problems of the Environment (SCOPE) and the International Council of Scientific Unions (ICSU) in this particular case.

It must not be assumed, however, that all is now known about the nature and causes of climatic change, and that only their impacts need to be assessed. The analysis and interpretation of actual changes in specific areas still needs considerable expansion (e.g. Flohn and Fantechi 1984), whilst the present state of knowledge (and gaps in such knowledge) of the processes involved and possible causes are reflected in a series of recent major books (e.g., Mörner and Karten 1984; Houghton 1984; Malone and Roederer 1985).

This relatively recent intensification of concern has derived from many pressures. Amongst these, three at least should be stressed. The first is that the impact of exceptional climatic conditions (droughts, floods, hurricanes, etc.) are now seen around the world as they happen, with television bringing them into the living room in all their stark drama. Thus there is a social awareness of climatic hazards across the globe which did not really exist previously. Scientifically, however, the combination of significant growth in the understanding of atmospheric processes and the development of sophisticated and powerful computer systems has facilitated the improvement of atmospheric mathematical modelling via global circulation models. These in turn not only have a vital role in short- to medium-term weather forecasting, but also present great potential for the simulation of past, present and future climates on the basis of postulating a variety of surface and atmospheric conditions. The symbiosis of meteorology and climatology is best reflected in the three streams of the World Climate Research Programme (WCRP) established in 1979 under the joint aegis of ICSU and the World Meteorological Organisation (WMO) as a component of the overall World Climate Programme. The objects of this have been described by its Director (Morel 1985) as being to obtain a 'better understanding of climate change and variability and their causes, whether from natural or human influences'. The first stream of work aims at establishing the physical basis for the prediction of weather anomalies on time-scales of one or two months; the second aims at predicting variations of global climate over periods of up to several years; and the third is trying to characterise climatic variations over periods of several decades, and to assess the potential response of climate to either natural or man-made influences (see also Houghton and Morel 1984).

It is the problem of the possible effects that humans, in their present societies and economies, may be having upon global climates that forms the third pressure leading to a growth in the 'climatic change industry'. The consumption of fossil fuels to sustain and expand industrial production and living standards in a world with a rapidly growing population, has led to a release into the atmosphere not only of increasing amounts of carbon dioxide but also a wide range of other gaseous components. Individually and collectively these may be approaching the point where the earth's natural physical-biological system cannot cope with radical changes occurring in the global climate. These tendencies are intensified by the dramatic changes in land use and vegetation cover that are being produced by the perceived need for more new farming land and agricultural output. Realising that if fundamental climate changes do occur for these reasons they may well prove to be irreversible, scientists are clearly concerned both to define the dimensions of the problem, and to suggest measures necessary to avoid it, before it is too late. Within the numerous recent international volumes mentioned earlier, as well as in the research literature as a whole, this theme bulks large.

Where does the present volume fit into this spectrum of studies? It derives from one subset of this multi-disciplinary study of climatic change (Yeh and Fu 1985), namely the Study Group on Recent Climatic Change established by the International Geographical Union in 1984. The papers were presented at a symposium held at the University of Sheffield, England, in August 1987, at which participants were not only geographer climatologists but also climatological colleagues from other allied disciplines. Collectively the papers reflect the wide range of studies that comprise the field of climatic change, from problems of deriving a reliable data base to attempts at causal explanation via mathematical modelling or atmospheric circulation studies; from description of past changes in statistical terms to probability estimates of the future, and forecasting by a mixture of statistical and atmospheric process approximations; and ultimately to evaluations of the implications of present and future changes for our society.

To organise their presentation in these terms is not easy, however, for so many contributions overlap several of these, and other, themes. Perhaps their major characteristic is that, apart from a few globally-orientated contributions, most arguments are presented in the context of one specific area of the earth's surface. This may not be surprising when the majority of contributors are geographers, for whom such an approach is 'natural'. Moreover, it complements rather than repeats the approach that is fundamental in the WCRP, with its concentration on global rather than regional or local trends and relationships. Once that programme has achieved operational models that are fully acceptable at the global or macro-level, there will still be the need to modulate them, and indeed to test them, in terms of the known climate changes of specific locations or areas. Moreover, until those models are fully established, area-specific studies will still provide the most useful input for both assessing applied impact and evaluating policy issues.

References

Ahlmann, H.W. 1948. 'The present climatic fluctuation', *Geog. J.* 112, 165–93.

Ahlmann, H.W. 1953. 'Glacier variations and climatic fluctuations', Bowman Memorial Lectures, *Amerc. Geogr. Soc.*, ser. 3, New York.

Bowman, I. 1909.'Man and climate change in South America', *Geog. J.* 33, 267–78.

Bowman, I. 1935. 'Our expanding and contracting "Desert"', *Geog. Rev.* 25, 43–61.

Brooks, C.E.P. 1928. 'Historical climatology of England and Wales', *Q.J.R. Met. Soc.* 54, 309–17.

Brooks, C.E.P. 1931. 'Changes of climate in the Old World during historic times', *Q.J.R. Met. Soc.* 57, 13–26.

Flohn, H., Fantechi, R. (eds) 1984. *The Climate of Europe: Past, Present and Future*, Dordrecht, D. Reidel.

Gregory, J.W. 1914. 'Is the Earth drying up?', *Geog. J.* 43, 148–72, 293–313.

Houghton, J.T. (ed.) 1984. *The Global Climate*, Cambridge, Cambridge University Press.

Houghton, J.T., Morel, P. 1984. 'The World Climate Research Programme' in J.T. Houghton (ed.), *The Global Climate*, Cambridge, Cambridge University Press, 1–11.

Huntington, E. 1907. *The Pulse of Asia*, Boston, Houghton Miflin & Co.

Huntington, E. 1912. 'The fluctuating climate of North America', *Geog. J.* 40, 264–80.

Huntington, E. 1914. 'Climatic changes', *Geog. J.* 44, 203–10.

Huntington, E. 1915. *Civilisation and Climate*, New York, Yale University Press.

Huntington, E. 1916. 'Climatic variations and economic cycles', *Geog. Rev.* 1, 192–202.

Kates, R.W., Ausubel, J.H., Berberian, M. (eds) 1985. *Climate Impact Assessment*, Chichester, John Wiley and Sons.

Kropotkin, Prince. 1904. 'The desiccation of Euro-Asia', *Geog. J.* 23, 722–34.

Ladurie, E. Le Roy 1971. *Times of Feast, Times of Famine: a History of Climate since the year 1000* (trans. B. Bray), New York, Doubleday.

Lamb, H.H. 1977. *Climate: Present, Past and Future — Volume 2: Climatic History and the Future*, London, Methuen.

Lamb, H.H. 1982. *Climate, History and the Modern World*, London, Methuen.

Malone, T.F., Roederer, J.G. (eds) 1985. *Global Change*, Cambridge, Cambridge University Press.

Manley, G. 1944. 'Some recent contributions to the study of climatic change', *Q.J.R. Met. Soc.* 70, 197–219.

Manley, G. 1949. 'The extent of the fluctuations shown during the "instrumental" period in relation to post-glacial events in N.W. Europe', *Q.J.R. Met. Soc.* 75, 165–71.

Manley, G. 1950. 'On British climatic fluctuations since Queen Elizabeth's day', *Weather* 5, 312–18.

Manley, G. 1951. 'The range of variation of the British climate. Part I: climatic fluctuations in modern times', *Geog. J.* 117, 43–65.

Manley, G. 1953. 'The mean temperature of Central England, 1698–1952', *Q.J.R. Met. Soc.* 79, 242–61.

Manley, G. 1959, 'Temperature trends in England, 1698–1957', *Arch. Met. Geoph. Biocl.* ser. B., 9, 413–33.

Manley, G. 1974. 'Central England temperatures: monthly means 1659–1973', *Q.J.R. Met. Soc.* 100, 389–405.

Morel, P. 1985. 'The World Climate Research Programme' in T.F. Malone and J.G. Roederer (eds), *Global Change*, Cambridge, Cambridge University Press, 171–81.

Mörner, N.A., Karten, W. (eds) 1984. *Climate Changes on a Yearly to Millennial Basis*, Dordrecht, D. Reidel.

Schouw, J.F. 1828 'On the supposed changes in the Meteorological Constitution of the different parts of the Earth during the Historical Period', *Edinb. Journ. Sci.* 8, 311–26.

Wigley, T.M.L., Ingram, M.J., Farmer, G. (eds) 1981. *Climate and History: Studies in Past Climates and their Impact on Man*, Cambridge, Cambridge University Press.

Yeh, T. Fu, C. 1985. 'Climate change — a global and multidisciplinary theme' in T.F. Malone and J.G. Roederer (eds), *Global Change*, Cambridge, Cambridge University Press, 127–45.

Part I
Global View

Chapter 1

Modelling the climatic response to greenhouse gases

W. Bach University of Münster
Federal Republic of Germany

The greenhouse problem

The Earth's climate depends on the incoming solar radiation and its absorption at the surface and in the atmosphere. The absorption within the atmosphere, in turn, depends on the types of gases and their concentrations. If we alter these, for example through the emission of additional radiatively active trace gases, we must realise that this will also alter the climate. Since the warming effect of the trace gases is somewhat similar to that in a greenhouse, they have also become known as greenhouse gases (GHG).

Among the presently most important GHG belong carbon dioxide (CO_2), methane (CH_4), nitrous oxide (N_2O), chlorofluorocarbons (such as CFC-11 and CFC-12), and ozone (O_3). So far, more than 30 additional GHG have been identified (Ramanathan *et al*. 1985) and the list is still growing. The most important anthropogenic sources of these GHG are the burning of fossil-fuels such as coal, oil, and gas in power stations, industry and homes as well as petrol and diesel in automobiles; agricultural activities including cultivation of rice, cattle farms and artificial fertilisation; the production and use of synthetic chemicals; deforestation in the tropics; forest dieback in mid- and high latitudes; and soil destruction worldwide. There is a growing scientific consensus that the combined effect of these sources will lead to an unprecedented climatic change within the next decades with potentially far-reaching impacts on ecosystems, food production, water resources, sea level, etc. In a world with a rapid population increase, this is of more than academic interest (Bach 1984).

Some climatic change impacts are inevitable due to man's past actions, and more is in store as a result of ongoing activities. A wait-and-see policy which hopes that mankind will be able to adapt to the changes is one of high risk, especially to the majority of the people in the Third World. A safer strategy would be to follow the WMO (1980) appeal to nations, namely 'to foresee and prevent potential man-made changes in climate that might be adverse to the wellbeing of humanity'.

Decisions based on such a safety struggle require information on uncertain events in the future. The appropriate methods for acquiring such information are climate modelling and scenario analysis. Specifically, this chapter illustrates the use of these methods, one important purpose of which is to make transparent to decision-makers the effectiveness of the various measures available to minimise climatic risks.

The modelling concept

The modelling concept used here is outlined in Figure 1.1. Carbon dioxide emission rates and CFC emission rates are calculated from various energy

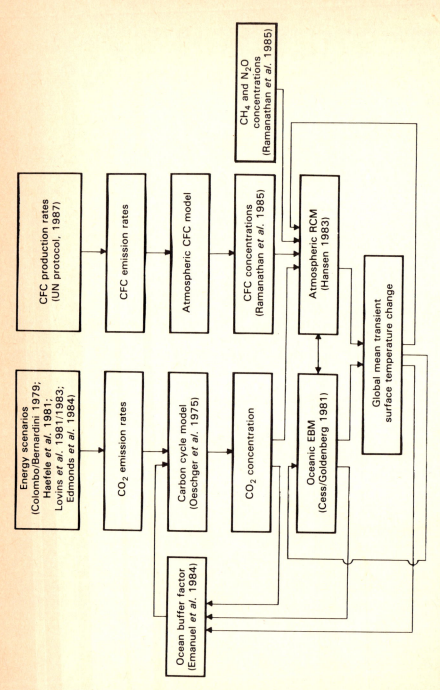

Figure 1.1 Modelling concepts

scenarios and CFC production rates, respectively. They serve as input in a carbon cycle model and a CFC mass balance model to compute the respective CO_2 and CFC concentrations. Together with historic and projected CH_4 and N_2O concentrations they serve as input for an atmospheric radiative-convective model (RCM) coupled with an oceanic energy balance model (EBM) to calculate the global mean transient surface temperature change.

Energy scenarios

Future world energy is one of the major uncertainties. Therefore, a set of five energy scenarios is used which is widely considered to span the probable uncertainty range from 1980 to 2100.

Oak Ridge A, as developed by Edmonds *et al.* (1984), projects energy demand to 2075. For the projection to 2100 it is considered reasonable to let coal increase linearly while oil and gas increase exponentially. The cumulative fossil-fuel use reaches 7450 TW in 2100, exhausting the probably technically recoverable fossil-fuel reserves of 3084 TW as early as 2065. This is an unrealistic scenario, unless one assumes the availability of an unlimited energy source that can be tapped at affordable cost. The present global primary energy use is approximately 10 TW.

The two scenarios from the International Institute for Applied Systems Analysis (IIASA), a high and a lower one as developed by Haefele *et al.* (1981), are included in this study because they are widely quoted, despite their considerable methodological flaws (see e.g., Keepin and Wynne 1984). Energy demand projections are from 1975 to 2030. Linear extrapolation to 2100 is done on the basis of the growth rate of 2020–30. The cumulative fossil-fuel use in 2100 reaches 3447 TW and 1949 TW, respectively, exhausting the probably technically recoverable fossil-fuel reserves in the high IIASA scenario shortly before the end of the twenty-first century in 2093.

A 'zero-growth' scenario was developed by Colombia and Bernardini (1979) for the EC assuming a constant per capita energy use and a doubling of world population thereby resulting also in a double edged energy demand in 2030. Linear extrapolation to 2100 results in a cumulative fossil-fuel use of 1296 TW which would not exhaust the probably technically recoverable fossil-fuel reserves by then.

The efficiency scenario developed by Lovins *et al.* (1981) for the German government projects energy demand to 2030. It follows a least-cost strategy implying that over the next 50 years all those efficiency improvements will be carried out which use presently available cost-effective technology. The efficiency energy use results in a fossil-fuel reduction of approx. 3.2 per cent per annum for the period 2020–30 which is linearly extrapolated to 2100. The cumulative fossil-fuel use reaches only some 250 TW in 2100, thereby stretching significantly the presently known fossil-fuel reserves.

The C-emissions are calculated from UN statistics for 1860–1980 and from the specifics of the scenarios from 1980 to 2100. Figure 1.2 shows that they range from about 5 Mt to more than 111 Gt. This gives a very large uncertainty range which in 2100 ranges from 1/1000 to 22 times the 1980 values of 4.96 GtC.

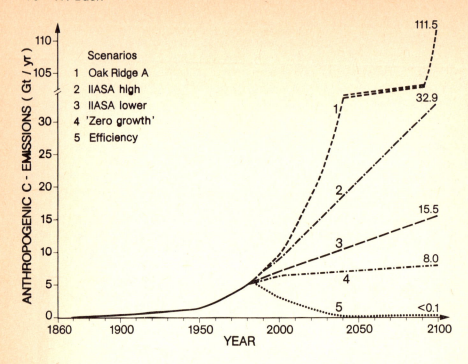

Figure 1.2   Anthropogenic C-emissions for a variety of energy scenarios

Carbon cycle model

The calculated CO_2 emissions from fossil-fuel sources are used to compute the historic and future atmospheric CO_2 concentrations. A box-diffusion carbon cycle model is used as developed by Oeschger *et al.* (1975) which consists of four reservoirs acting both as sources and sinks, namely the atmosphere, the biosphere, a well-mixed ocean layer (approx. 75 m deep), and the deep sea (approx. 4000 m deep) subdivided into 42 layers in which transport is by eddy diffusion. This model reproduces satisfactorily the measured ^{14}C profiles in the ocean. The ocean chemistry which determines the efficiency of oceanic CO_2 uptake is calculated explicitly for each time step as a function of temperature as well as oceanic and atmospheric CO_2 concentration using the Oak Ridge carbon cycle model (Emanuel *et al.* 1984). As shown in Figure 1.1 this allows to calculate a time-dependent buffer factor which is used in the Oeschger *et al.* model. The resulting atmospheric CO_2 concentrations for 2100 span the large uncertainty range of 367 ppm for the efficiency scenario to close to 2500 ppm for scenario Oak Ridge A which is 1.1 to 7.2 times the present concentration of 345 ppm.

Trace gas scenarios

Four scenarios are considered for the trace gases CH_4, N_2O, CFC-11 ($CFCl_3$) and CFC-12 (CF_2Cl_2) to span the probable uncertainty range from 1980 to 2100.

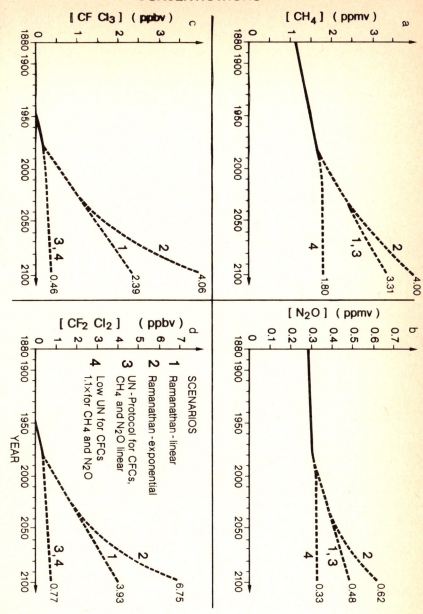

CONCENTRATIONS

Figure 1.3 Scenarios of the trace gas concentrations

The 'Ramanathan linear' scenario (no 1. in Figure 1.3) uses the trace gas concentration changes as given by Ramanathan *et al.* (1985) from 1880 to 2030 and extrapolates linearly to 2100 based on the 2020–30 concentration change. This gives in 2100 values of 3.31 ppm for CH_4, 0.48 ppm for N_2O, 2.39 ppb for CFC-11, and 3.93 ppb for CFC-12, which exceed the 1980 concentrations by factors of 2, 1.6, 13 and 14, respectively.

The 'Ramanathan exponential' scenario (no. 2. in Figure 1.3) is the same as no. 1 except that it extrapolates exponentially from 2030 to 2100. This results in values in 2100 of 4.00 ppm for CH_4, 0.62 ppm for N_2O, 4.06 ppb for CFC-11, and 6.75 ppb for CFC-12, which exceeds the respective 1980 trace gas concentrations by factors of 2.4, 2.1, 23 and 24, respectively.

The 'UN Protocol' scenario (no. 3 in Figure 1.3) is based on the following agreement reached at the September 1987 UN Conference in Montreal. A 1990 freeze of the CFC-11 and CFC-12 production rates at 1986 levels, a 20 per cent production cut in 1992 and the option (here considered as realised) of an additional 30 per cent cut by 2000. Calculations are carried out with a simple CFC mass balance model as described below. These measures lead to much reduced CFC-11 and CFC-12 concentrations which, in 2100, reach respectively 0.46 ppb and 0.77 ppb, which is only 2.6 and 2.8 times the 1980 values. The CH_4 and N_2O are those used in scenario no 1.

The 'low' scenario (no. 4 in Figure 1.3) is identical to the 'UN Protocol' scenario except that the CH_4 and N_2O concentrations are asymptotically increased by 1.1 times the 1980 value until 2100. This trace gas scenario is developed to be used in conjunction with the energy efficiency scenario for CO_2. The combination of the low trace gas scenario and the efficiency energy scenario is subsequently called the 'toleration' scenario which, if carried through, would minimise the anthropogenic effects on climate.

Mass balance model for CFC

Global CFC production rates are taken from Hammit *et al.* (1986) and reduced according to the UN protocol agreement. Of the produced CFCs only 70 per cent are emitted into the atmosphere in the first year, the rest is released in an exponentially decreasing quantity. The CFC mass balance in the troposphere and the stratosphere is assumed to be 90:10. In the stratosphere photolysis is described as the exponential decrease of an added CFC quantity. The concentrations by volume in the respective atmospheres are calculated from the molecular weights, the shares of the various masses, and the total mass of the atmosphere.

Climate model

The calculated trace gas changes are used as input in a climate model (Figure 1.1) consisting of the parameterised form of a one-dimensional radiative-convective model (RCM) of the atmosphere developed by Hansen (1983) and a one-dimensional energy balance model of the ocean of a type similar to that used by Cess and Goldenberg (1981) to calculate the global mean transient surface temperature change from 1860 to 2100. The RCM considers the concentration changes of CO_2, CH_4, N_2O, CFC-11 and CFC-12; as well as changes in volcanic activity, the 'solar constant', and surface temperature. It does not take into account concentration changes of O_3 and additional CFCs nor the rather complex chemical reactions of the ozone photochemistry.

The EBM calculates the heat flux in the ocean. For that purpose the ocean is subdivided into 42 vertical layers — analogously to the carbon cycle model. The ocean mixed layer (about 75 m) is assumed to be fully mixed. In the deep sea (about 4000 m) energy transport is modelled as a vertical diffusion process. Of the many possible feedback processes only the ocean buffer factor is explicitly taken into account (see Figure 1.1). Other feedbacks are included in parameterised form.

Modelling results

Temperature change due to CO_2 only

Figure 1.4 shows the calculated transient global mean surface temperature change due to CO_2 only for 1860–2100 using the above climate model. The calculations for 1860–1980 are based on historical carbon data, whereas the projections to 2100 are based on the indicated energy scenarios spanning the likely range of future development. The vertical scales permit the surface temperature changes ΔT_s to be related to the most commonly used reference periods of 1860, 1946–60, and 1980. Reference is also made to paleoclimatic warm periods (related here to the middle period of 1946–60) to give an indication of the severity of the changes. In the Oak

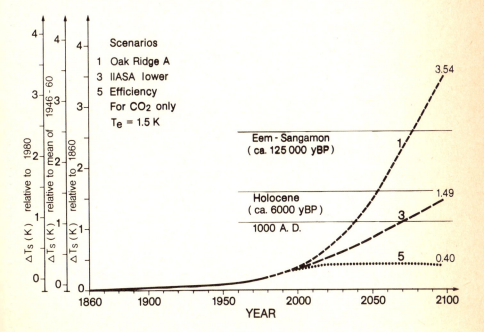

Figure 1.4 Calculated transient global mean surface temperature change, ΔT_s (K), due to CO_2 only and for the low climate sensitivity of $T_e = 1.5$ K

Ridge scenario A the warming in 2100 could be higher than that of the Eem-Sangamon Interglacial.

The three-dimensional general circulation models, which are presently our best available models, calculate for a doubling of CO_2 a new equilibrium temperature of $T_e = 1.5-4.5$ K (NRC 1983). This temperature range is called the climate sensitivity and it reflects the present uncertainty in climate modelling. Taking this uncertainty range into account, the calculated ΔT_s shown in Figure 1.4 range from 0.40 to 3.54 K in 2100 relative to 1860 for $T_e = 1.5$ K; for $T_e = 4.5$ K the range is from 0.94 to 6.40 K. It is therefore not uncritical for which climate sensitivity the calculations are made.

Temperature change due to CO_2 and the other greenhouse gases

Similar transient temperature-change estimates are made for the combined effects of CO_2 and the other greenhouse gases such as CH_4, N_2O, CFC-11 and CFC-12. Calculations in Figure 1.5 are for $T_e = 1.5$ K and in Figure 1.6 for $T_e = 4.5$ K. The projections reflect the respective energy and trace gas scenarios. For $T_e = 1.5$ K ΔT_s ranges from 0.90 to 4.51 K in 2100 relative to 1860; and for $T_e = 4.5$ K the range is even as large as 2.05–8.33 K.

A study group of the WMO (1985) obtained from their assessment of the same types of trace gases as used in this study a cumulative equilibrium surface warming in the range of 1.5–6.1 K from 1850 to 2030. Mintzer (1987), including also ozone

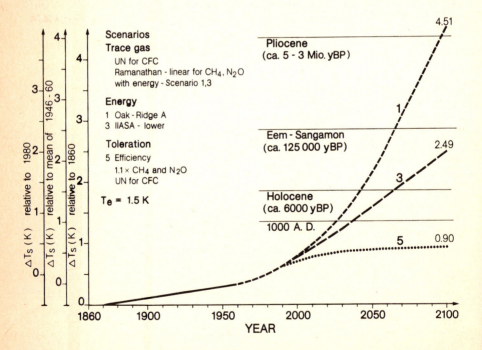

Figure 1.5 Calculated transient global mean surface temperature change, ΔT_s(K), due to CO_2 and the other greenhouse gases (CH_4, N_2O, CFC 11, CFC–12) and for the low climate sensitivity of $T_e = 1.5$ K

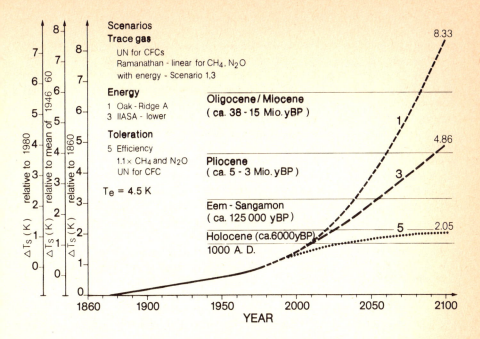

Figure 1.6 Calculated transient global mean surface temperature change, $\Delta T_s(K)$, due to CO_2 and the other greenhouse gases (CH_4, N_2O, CFC–11, CFC–12) and for the high climate sensitivity of $T_e = 4.5$ K

forcing, obtained by a scaling procedure an equilibrium warming commitment in 2075 relative to the pre-industrial atmosphere of 1.4–4.2 K, 2.3–7.0 K, 2.9–8.6 K, and 5.3–16.0 K for the 'slow build-up', 'modest policies', 'base case', and 'high emissions' scenarios respectively.

Policy implications

Should the above temperature projections come about over the next century, this would indeed be a frightening prospect, considering that they are global mean values. Depending on the region, they should be amplified by a factor of up to two or three to allow for the relatively greater warming in polar regions. All of these values go beyond what mankind has ever experienced before, and those in the higher range exceed even the temperature of the Oligocene/Miocene epoch, which, some 38 to 15 million years before the present, was with approx. 6 K the warmest period that can be somewhat reliably constructed from the palaeoclimatic record (Figure 1.6). If society decides that these prospects are unacceptable, decision-makers will be called upon for co-ordinated action. The necessary information on which to base rational decisions must come from the scientist.

'Toleration' scenario

The decision-maker needs information on how great a climatic change (temperature increase) can be tolerated, or, if one takes past actions into account, must be tolerated. In a joint statement the German Meteorological Society and the German Physical Society have concluded that, according to present knowledge, the global mean surface temperature should not rise by more than 1 K above the current level, because even such a seemingly modest increase in temperature could have noticeable impacts on the regional climate (DMG/DPG 1987). An increase of 1 K would limit the addition of greenhouse gases to the atmosphere to a CO_2 equivalent concentration of approx. 450 ppm. Assuming that the contribution of the other greenhouse gases equals that of CO_2, then the present CO_2 content of the atmosphere of 345 ppm could only be permitted to increase by another 55 ppm. At the current airborne fraction of approx. 50 per cent, this would correspond to a fossil-fuel use of about 240 TW. Thus, at the current rate, this would permit fossil-fuel combustion for only another 30 years.

What combination of scenarios would occur with a temperature limit of about 1 K or 1.5 K above the present or the pre-industrial levels, respectively? Our model runs show that a combination of the energy efficiency scenario with the low trace gas scenario would result in a global mean surface temperature increase of $\Delta T_s = 0.90$ K and $\Delta T_s = 2.05$ K relative to 1860 for $T_e = 1.5$ K and $T_e = 4.5$ K respectively. Some consider the value for the lower climate sensitivity of $T_e = 1.5$ K to be the most realistic because in 1980 it yields a temperature rise of 0.51 K which comes very close to the actually measured global mean temperature increase of 0.4–0.6 K over the past century (Jones et al. 1986). Others argue that the most sophisticated general circulation models with their climate sensitivity closer to $T_e = 4.5$ K give better estimates of a future temperature change, although in 1980 this yields a temperature increase of approx. 0.95 K which exceeds that measured considerably. To allow for this uncertainty it is best to use the range of 1–2 K for the temperature increase by 2100. The combination of efficiency and low trace gas scenarios required to stay within this limit has been dubbed the 'toleration' scenario, because the trace gases have been reduced as much as is probably politically possible and the resulting warming may just have to be tolerated.

Required emission reduction

A final step leads to the most painful part of the decision-making process, namely the allocation of the individual emission reductions. The usual approach is an across-the-board percentage cut. While this often gives the impression of fairness, it makes neither economic sense nor does it help solve the climate-change problem. A more rational approach is to calculate first, with the help of a climate model, the contribution of each greenhouse gas to a warming limit, which can be or must be tolerated, and then to decide by how much each individual source or use can be reduced to meet the required emission reduction for each gas.

The rationality of this approach becomes apparent when examining the details given in Table 1.1. For example, with increasing population, especially in the developing countries, it would not be realistic to expect much of a reduction in CH_4 from rice paddy fields or in N_2O from the cultivation of land. Therefore, in the last column, the annual reduction from these sources has been set close to zero.

Table 1.1 Estimates of the required trace gas emission reduction by use to meet the global mean temperature increase limit of 1–2 K in 2100 relative to 1860 for the climate sensitivity range of T_e 3 ± 1.5 K

Gas	Source/use	Emission rate c.1980 (t)		Required reduction until 2100 to (t)	Annual reduction beginning 1980 (%/yr)
		$\times 10^6$		$\times 10^6$	
	Coal	1862		1.75	5.6
	Oil	2369		0.14	7.8
CO_2	Gas	726	1)	1.07	5.3
	Misc.	–		–	
	Total	4957		2.97	6.0^5
		$\times 10^6$		$\times 10^6$	
	Paddy fields	69–167		27–167	0.8–0.0
	Ruminants	60– 99		3– 10	2.5–1.9
CH_4	Biomass burning	56–102		5– 20	2.0–1.3
	Leakage of natural gas	30– 40	2)	1– 1	2.8–3.0
	Coal mining	30– 40		1– 5	2.8–1.7
	Misc.	55–102		1– 30	3.3–1.0
	Total	300–550		38–233	1.7–0.7
		$\times 10^6$		$\times 10^6$	
	Fertiliser	0.6–3.0		0.1–0.2	1.5–2.2
	Cultivation of land	1.5–3.0		1.6–3.0	0.0–0.0
N_2O	Fossil fuel burning	1.8–1.9	3)	0.3–0.4	1.5–1.3
	Biomass burning	1.0–2.0		0.2–0.4	1.3–1.3
	Misc.	4.4–5.1		0.8–1.0	1.4–1.3
	Total	10.0–15.0		3.0–5.0	1.0–0.9
		$\times 10^3$		$\times 10^3$	
	Foam	174		35	1.3
	Aerosol	93		5	3.1
$CFCl_3$	Refr. + air cond.	9	4)	5	0.6
	Misc.	24		18	0.03
	Total	300		63	1.3
		$\times 10^3$		$\times 10^3$	
	Aerosol	116.8		6	2.4
	Refr. + air cond.	98.5		25	1.1
CF_2Cl_2	Foam	43.8	4)	18	0.7
	Misc.	105.9		40	0.8
	Total	365.0		89	1.2

Notes:　1　Lovins *et al.* (1981)
　　　　2　Seiler (1984; 1985a); Crutzen (1986, 1987)
　　　　3　Crutzen (1983); Seiler (1985b)
　　　　4　Hammit *et al.* (1986) for 1984
　　　　5　Slow reduction period after 1980, fast reduction period toward 2100 (average reduction approx. 3.2 per cent per annum)

On the other hand, the very large conservative potential in fossil-fuel use would permit a significant reduction in CO_2 emission. Similarly, the reduction potential for $CFCl_3$ and CF_2Cl_2 is very large. Their production for use as aerosols in spray cans, foams and coolants can be drastically reduced, since they can be recovered, recycled or substituted at acceptable cost (Miller and Mintzer, 1986).

No doubt there will be different perceptions on the individual reduction potential. Any reallocation between the individual sources or uses is welcome provided this can be substituted by argument and it adds up to the required total reduction for each gas. To meet a global mean temperature increase limit of 1–2 K in 2100 relative to 1860, this study finds that from now on this would require an annual emission reduction of approx. 3.2, 1.7–0.7, 1.0–0.9, 1.3 and 1.2 per cent for CO_2, CH_4, N_2O, $CFCl_3$ and CF_2Cl_2 respectively.

References

Bach, W. 1984. *Our Threatened Climate: Ways of Averting the CO_2 Problem through Rational Energy Use*, Dordrecht, D Reide.

Cess, R.D., Goldenberg, S.D. 1981. 'The effect of ocean heat capacity upon global warming due to increasing atmospheric carbon dioxide', *J. Geophys. Res.* 86, 498–502.

Colombo, U., Bernardini, D. 1979. *A Low Energy Growth 2030 Scenario and the Perspectives for Western Europe*, Report prepared for the Commission of the European Communities, Brussels.

Crutzen, P.J. 1983. 'Atmospheric interactions — homogeneous gas reactions of C, N, and S containing compounds' (In B. Bolin and R.B. Cook (eds) *The Major Biochemical Cycles and their Interactions*, SCOPE 21, Wiley, Chichester), 67–112.

Crutzen, P.J. 1986. 'Globale Aspekte der atmosphärischen Chemie: Natürliche und anthropogene Einflüsse', Vorträge N 347, Rheinisch-Westfälische Akademie der Wissenschaften, Opladen, Westdeutscher Verlag, 41–72.

Crutzen, P.J. 1987. 'Role of tropics in atmospheric chemistry' (In R.E. Dickinson (ed.), *The Geophysiology of Amazonia. Vegetation and Climate Interactions*, Wiley, Chichester), 107–31.

DMG/DPG (Deutsche Meteorologische Gesellschaft/Deutsche Physikalische Gesellschaft) 1987. *Warnung vor drohenden weltweiten Klimaänderungen durch den Menschen*, Bad Honnef.

Edmonds, J.A., Reilly, J., Trabalka, J.R., Reichle, D.E. 1984. *An Analysis of Possible Future Atmospheric Retention of Fossil Fuel CO_2*, Dept. of Energy, TR 013, Washington, DC.

Emanuel, W.R., Killough, G.G., Post, W.M., Shugart, H.H., Stevenson, M.P. 1984 *Computer Implementation of a Globally Averaged Model of the World Carbon Cycle*, Dept. of Energy, TR 010, Washington, DC.

Haefele, W. *et al.* 1981. *Energy in a Finite World*, Vols. 1 and 2. Ballinger, Cambridge, MA.

Hammit, J.K. *et al.* 1986. *Product Uses and Market Trends for Potential Ozone-depleting Substances, 1985–2000*, report prepared for the US EPA by the Rand Corporation, Santa Monica, CA.

Hansen, J. 1983. Cited in Seidel, S. and Keyes, D., *Can We Delay a Greenhouse Warming?* Washington DC, US EPA.

Jones, P.D., Wigley, T.M.L., Wright, P.B. 1986. 'Global temperature variations between 1861 and 1984', *Nature* 322, 430–4.

Keepin, B., Wynne, B. 1984. 'Technical analysis of IIASA energy scenarios', *Nature* 312, 691–5.

Lovins, A.B., Lovins, L.H., Krause, F., Bach, W. 1981. *Least-cost Energy: Solving the*

CO₂ Problem, Andover, MA, Brick House.

Miller, A.S., Mintzer, I.M. 1986. *The Sky is the Limit: Strategies for Protecting the Ozone Layer*, Research Report no. 3, World Resources Institute, Washington, DC

Mintzer, I.M. 1987. *A Matter of Degrees: The Potential for Controlling the Greenhouse Effect* Research Report no. 5, World Resources Institute, Washington, DC.

NRC (National Research Council) 1983. *Changing Climate*, National Academy Press, Washington, DC.

Oeschger, H., Siegenthaler, U. Schatterer, U. and Gugelmann, A. 1975. 'A box diffusion model to study the carbon dioxide exchange in nature', *Tellus* 27, 168–92.

Ramanathan, V., Cicerone, R.J. Singh, H. B., Kiehl, J.T. 1985. 'Trace gas trends and their potential role in climatic change, *J. Geophys. Res.* 90(D3), 5547–66.

Seiler, W. 1984. 'Contribution of biological processes to the global budget of CH₄ in the atmosphere' (In M.J. Klug and C.A. Reddy (eds) *Current Perspectives in Microbial Ecology*), Amer. Soc. Microbiology, Washington, DC.

Seiler, W. 1985a. *Increase of Atmospheric Methane: Causes and Impact on the Environment*. Special Environmental Report no. 16, WMO, Geneva.

Seiler, W. 1985b. 'Cycles of radiatively important trace gases (CH₄, N₂O)' in *The Impact of an Increased Atmospheric Concentration of Carbon Dioxide on the Environment*, Geneva, WMO/ICSU/UNEP.

WMO. 1980. *Outline Plan and Basis for the World Climate Programme 1980–1983*, WMO no. 540, Geneva.

WMO. 1985. *Atmospheric Ozone 1985, Vol. III, Global Ozone Research and Monitoring Project*, report no. 16. Geneva.

Chapter 2

Volcanism and air temperature variations in recent centuries

C.-D. Schönwiese University of Frankfurt,
Federal Republic of Germany

Introduction

The ejecta of explosive volcanism, including gas-to-particle conversions, form stratospheric aerosol layers which influence the atmospheric radiation processes and, in consequence, the climate. Figure 2.1 reflects the LIDAR (light detection and ranging) measurements of the integral stratospheric particulate backscattering at the Garmisch-Partenkirchen station (FRG); data and analysis are taken from Reiter and Jäger (1986). These monthly data indicate not only some particular volcanic eruptions (St Helens, Alaid, Pagan, Chichón, Nevado del Ruiz) but also a pronounced long-term trend culminating in 1982–3. This long-term trend can be explained by the atmospheric residence time of the sulphuric acid aerosols, approximately one to five years (Hammer *et al.* 1981; Rampino and Self 1984; Jäger *et al.* 1984) and is important in the context of a climatological interpretation.

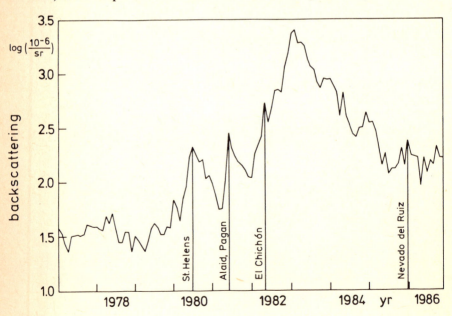

Figure 2.1 LIDAR measurements of the integrated particulate backscattering of the stratosphere

Source: Monthly mean data, Reiter and Jäger (1986), redrawn and some volanic eruptions indicated (logarithmic ordinate scale).

The volcanogenic aerosol layers (stratosphere) scatter the solar irradiation and only a reduced part of the multi-scattered irradiation can reach the earth's surface. So, the stratosphere may be warmed (increased absorption) whereas the decrease of the atmospheric transmission should lead to a cooling of the atmospheric boundary layer near the surface. Moreover, the climate response to explosive volcanic activity should be large-scale in space because the stratospheric sulphuric acid aerosols, due to their long residence time, undergo a widespread mixing (McCormick et al. 1978). Any analysis of volcanism–climate relationships must account for these features in time and space.

The most certain climate response to volcanic activity is that of temperature. It was discussed in a number of both theoretical (e.g., Pollack et al. 1976; Bryson and Goodman 1980; Toon and Pollack 1980; Hansen et al. 1981) and statistical studies (e.g., Miles and Gildersleeves 1978; Sear and Kelly 1982; Angell and Korshover 1985; Schönwiese 1984; 1986). In the following, the statistical point of view is again emphasised.

Volcanic activity parameters

Lamb (1970; 1983) was the first to try to evaluate parameter time series which quantify the volcanic particle loading of the stratosphere in recent centuries. His 'dust veil index' (DVI) is available in terms of annual means for the northern hemisphere from 1500 (southern hemisphere from 1890) until 1983. Despite Lamb's careful work the DVI time series is, however, inhomogeneous in nature (due to different regression equations) and the DVI data, zero within the whole period 1916–62, seem to be unrealistic when compared with the Smithsonian volcano chronology (Simkin et al. 1981).

It is very helpful, therefore, that in addition to DVI, alternative series are available based on ice core measurements, in particular the Crête (central Greenland) measurements specifying the acidity of the ice core samples and reflecting, to a certain extent, the acid volcanogenic material deposited in the polar ice. This Crête 'acidity index' (AI) series was provided by Hammer et al. (1981) and is available, also in terms of annual data, for the period 553–1972. There are, however, again some shortcomings implied. First of all, the AI record is of regional type and not necessarily representative for the northern hemisphere. In addition, the AI series does not strictly discern between eruptions which have affected the stratosphere and those which have not. Finally, the acidity may be influenced by non-volcanic environment effects.

These shortcomings of DVI and AI favour the evaluation of a third volcanic activity parameter. This was done on the basis of the Smithsonian volcano chronology edited by Simkin et al. (1981; and unpublished supplements, 1985). This chronology classifies all volcanic eruptions known from historical and geological sources in respect of the column height and volume of their ejecta in terms of the 'volcanic explosivity index' (VEI) on a scale of zero to eight (for instance Tambora, 1815, VEI=7, Krakatoa, 1883, VEI=6; Agung, 1963, VEI=4; St. Helens, 1980, VEI=5; El Chichón, 1982, VEI=4). Note that this index does not represent annual assessments but a classification of the particular eruptions. Newhall and Self (1982) state that the set of eruptions which needs to be considered in climatological studies is probably the set with VEI $\geq$ 3, from 1755 to the present. Schönwiese (1986) has tried to evaluate an annual volcanic activity parameter, following these recommendations and using the simple formula

$$SVI = \sum_{i=1}^{n} 10^{VEI}$$

where all eruptions $VEI \geq 3$ are summarised for each year (n eruptions per particular year). Due to the definition of VEI (Simkin *et al.* 1981; Newhall and Self 1982) this definition means a weighting by the volume of the assessed ejecta and an accumulation throughout the year. For northern hemisphere assessments we used all volcanic eruptions from 90° N to 10° S. The abbreviation 'SVI' means 'Smithsonian volcanic index' referring to the data source. Recently, we tried to imply also the inter-annual residence time of the sulphuric acid material. By comparing the VEI (eruption) and AI (deposition) data an exponential function was evaluated describing the decrease of the stratospheric aerosol concentration following the major explosive eruptions (where only some outstanding eruptions like Tambora were taken into account). The outcome was a 'lagged' SVI called SVI*.

Statistical analysis

In the first step of a statistical analysis of volcanism–temperature relationships one may look on large-scale effects concerning simultaneously the temperature of the stratosphere and the atmospheric boundary layer near the surface. This is done in Figure 2.2 where the stratospheric data from Angell and Korshover (1984) (A/K), available since 1958 but less reliable than the Labitzke *et al.* (1986) data (L), are plotted. These stratospheric temperature data, annual and northern hemisphere averages, are compared with corresponding data for the atmosphere near the surface, provided by Jones (1985a). The volcanic parameter SVI (Figure 2.2 bottom) indicates some particular years where major explosive volcanic eruptions were observed. Note that the SVI parameter is in fair agreement with the back-scattering measurements (Figure 2.1) especially if the El Chichón eruption is corrected from VEI=4 to VEI=5. (It seems that the Smithsonian chronology has overestimated the St Helens and underestimated the El Chichón eruption. In the following, however, no 'corrections' of the underlying VEI data are made.) Some particular eruptions are fairly well produced by the SVI parameter, for example Bezymianny 1956, (Agung, 1963), Kelut and Awu 1966, Fernandina 1968, Tiatia and Fuego 1973–4, Augustine 1976, St Helens 1980 and El Chichón 1982. A simultaneous stratospheric warming and cooling near the surface is indicated especially in (1956) 1964 (one year after the Agung eruption), 1968, (1974), 1976 and 1982 but not in 1980 (St Helens). The DVI parameter indicates more or less only the Agung and El Chichón eruptions.

Figure 2.3 compares the mean northern hemisphere temperature variations, year-to-year anomalies and ten-year low-pass filtered fluctuations (data before 1851 mainly from Groveman and Landsberg 1979) with the corresponding data of the volcanic parameters DVI, AI, SVI (Schönwiese 1986) and SVI*. Table 2.1 specifies the corresponding relation coefficients (AI not used) where, in addition to the northern hemisphere mean, the following temperature records are implied: Arctic mean (data from Jones 1985b), Philadelphia (USA), central England and Hohenpeissenberg (FRG, mountain station). The most significant results are found

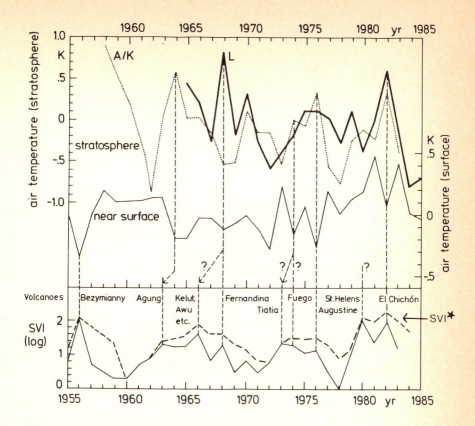

Figure 2.2 Upper plots of annual air temperature anomalies, northern hemisphere and annual averages

Sources: Stratospheric data 'A/K' 16–24 km averages from Angell and Korshover (1984), 'L' 24 km data from Labitzke *et al.* (1986); corresponding data near surface from Jones (1985a). This is compared, see lower plots, with corresponding data of the volcanic activity parameters SVI and SVI* (here corrected for El Chichón VEI=5; explanation see text, further volcanic parameters see Figure 2.3).

in relation to the Philadelphia record but in relation to the northern hemisphere mean the correlation coefficients also exceed the 95% confidence level. In general, the correlation coefficients increase from SVI to SVI* and it is speculated that the DVI parameter overestimates the relationships somewhat. The increase in case of using low-pass filtered data is in line with corresponding coherence analyses (Schönwiese 1983; 1984), indicating that the relationships are more established and also more confident in the long-term domain of the climate fluctuations. The low-pass filtered data simple correlation coefficients, however, fail to be significant because of considerable autocorrelation (mean Arctic data from Jones 1985b).

The regression equations enable the assessment of the temperature signals which may be hypothetically forced by volcanic activity. Table 2.2 lists the maximum signals simulating the temperature decrease due to the 'observed' minimum value ($\rightarrow$ temperature without volcanic influence $\rightarrow t_2$) and maximum value ($\rightarrow t_1$, signal

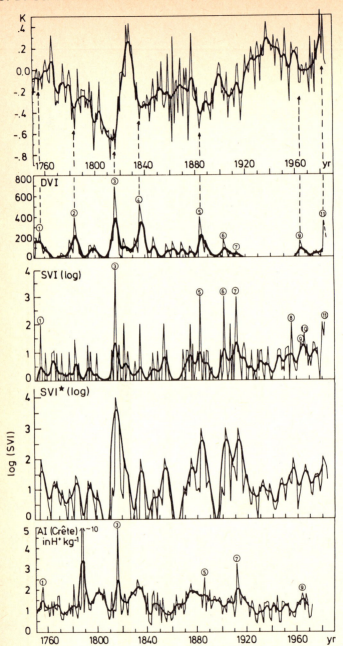

Figure 2.3 Mean northern hemisphere air temperature anomalies, annual and ten-year Gaussian low-pass filtered data

Sources: Deviations from the 1951–70 mean, before **1851**, mainly from Groveman and Landsberg 1979, later on from Jones 1985a. This is compared with corresponding data of the following volcanic parameter time series: dust veil index (DVI) from Lamb (1970; 1983), Smithsonian volanic index (SVI) and some index 'lagged' (SVI*) (explanation see text), and Greenland acidity index (AI) (ice core measurements) from Hammer *et al.* (1981).

Table 2.1 Correlation coefficients of some selected air temperature series (near surface) and volcanic parameter time series, annual and ten-year Gaussian low-pass filtered data

Record	Period	Annual data SVI	SVI*	DVI	Low-pass filtered SVI	SVI*	DVI
Northern hemisphere (mean)	1851–1984	−.17	−.26	−.27	−.34	−.40	−.62
	1781–1984	−.12	−.22	−.34	−.38	−.38	−.51
Arctic (mean)	1851–1984	−.18	−.21	−.28	−.34	−.37	−.64
Philadelphia (USA)	1781–1984	−.13	−.23	−.41	−.30	−.32	−.60
Central England	1781–1984	(−.02)	(−.09)	−.17	−.36	−.34	−.52
Hohenpeissenberg (FRG)	1781–1984	(−.06)	(−.11)	(−.12)	−.22	−.20	−.30

Note: For abbreviations and time series plots of the volcanic parameters see Figure 3 and text. The confidence tests take account of autocorrelation and non-Gaussian distributed samples. In case of the annual data analysis italic numbers exceed the 95% level. Numbers in parantheses do not exceed the 90% level. All low-pass filtered data correlations are not significant due to excessive autocorrelation which masks the relationships. In these cases a coherence analysis is needed (see Schönwiese 1983; 1984).

Table 2.2 Maximum temperature signals related to Table 2.1

Record	Period	Annual data SVI	SVI*	DVI	Low-pass filtered SVI	SVI*	DVI
Northern hemisphere (mean)	1851–1984	−.28	−.38	−.38	−.26	−.28	−.57
	1781–1984	−.47	−.53	−.62	−.70	−.64	−.57
Arctic (mean)	1851–1984	−.73	−.66	−.95	−.61	−.59	−1.37
Philadelphia (USA)	1781–1984	−.99	−1.53	−1.24	−.91	−.94	−.96
Central England	1781–1984	−.63	−1.00	−.43	−.69	−.57	−.56
Hohenpeissenberg	1781–1984	−.16	−.63	−.65	−.91	−.78	−.78

Note: Signals which exceed more than twice the annual data standard deviation are indicated by bold numbers.

$= t_1 − t_2$) of the volcanic parameters. Again in the case of Philadelphia the greatest numbers are calculated but the signals are considerable even in the case of non-significant correlation coefficients (but highly hypothetical in these cases).

In addition to spectral statistics not discussed here, moving techniques allow more insight in the volcanism–climate relationships. In Figure 2.4 moving correlation coefficients are plotted (100-year subintervals, moving in one-year steps) analysing the correlation of the northern hemisphere mean with the volcanic activity parameters DVI, AI and SVI (SVI* being very similar to SVI). This confirms the statement of Newhall and Self (1982) that the VEI data are uncertain before 1755 (Figure 2.4 indicates roughly the years 1750–70). It is an interesting result that not only the SVI but also the DVI and AI correlation coefficients show a sharp drop in direction to negative values in this time. Also interesting is the weak AI correlation until roughly 1870–80 (centre of subinterval), followed by a sharp increase of the correlation coefficient later on. Relatively stable in time are the SVI correlations whereas the DVI correlations emerge with the greatest numbers but this may be due to overestimations. Similarly, Figure 2.5 shows moving signals, but related to 30-year subintervals (moving again in one-year steps). It is clearly

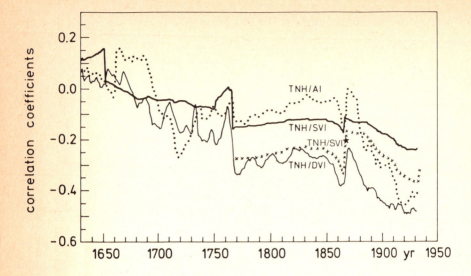

Figure 2.4 Moving correlation coefficients, 100 year subintervals and one year steps, plotted at the centre of each subinterval.

Note: Correlated are the mean northern hemisphere mean annual temperature data (TNH) (see Figure 2.3) with the volcanic activity parameters AI (dotted line), SVI (heavy solid line), SVI* (only since 1720–1819 subinterval, crosses), and DVI (lower solid line). Note that not before the subinterval centre *c*.1770 considerable correlations are indicated.

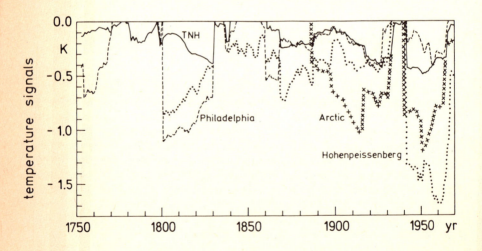

Figure 2.5 Moving temperature signals, 30 year subintervals and one-year steps, of the air temperature records indicated (annual means, TNH = northern hemisphere mean) as evaluated by a correlation analysis using the volcanic activity parameter SVI.

Note: This view reveals the effect (signal) of some outstanding volcanic eruptions (marked signal increase 15 years before those eruptions because of 30 year subinterval analysis), for example −1.1°C in case of Philadelphia and the Tambora eruption in 1815 (see also Tables 2.1 and 2.2).

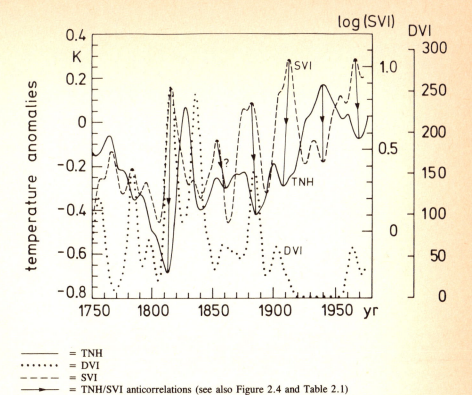

Figure 2.6 Mean northern hemisphere air temperature anomalies, 20-year low-pass filtered data TNH, compared with corresponding data of the volcanic activity parameter time series DVI and SVI

indicated that volcanism is very unstable in time and that the regional effects of volcanism are quite different. Philadelphia, for example, seems to be influenced predominantly by the Tambor eruption (temperature decrease indicated in Figure 2.5 c.15 years before this event because 30-year subintervals are used) whereas the Hohenpeissenberg and Arctic series reflect more intensively the Bezymianny and Agung eruptions (El Chichón is not considered here). In the long-term domain it is remarkable that SVI or SVI* are the only parameters which show an increase since c.1940 which, hypothetically, may be an explanation for the decrease of the northern hemisphere mean temperature since that time (see Figure 2.6). When compared with similar analyses of the anthropogenic greenhouse effect (Schönwiese 1987, Schönwiese and Malcher 1987) the volcanism–temperature correlations and signals, related to the northern hemisphere temperature fluctuations, are smaller ('greenhouse' signals, time series analysis from pre-industrial time until 1985, in the order of 0.7–0.9 °C, northern hemisphere). The regional effects of volcanism, however, seem to have been — so far — greater at some stations (regions). Figure 2.7 gives preliminary examples of statistical simulations of the northern hemisphere temperature fluctuations (ten-year low-pass filtered) on the base of multivariate analyses implying volcanic, solar and CO_2 forcing.

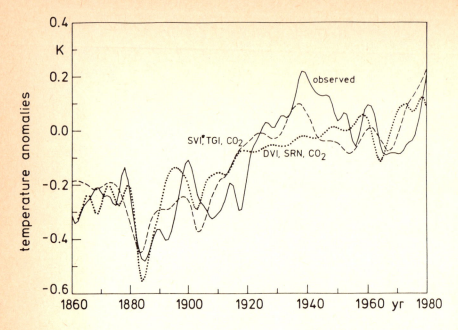

Figure 2.7   Some preliminary results simulating the northern hemisphere air temperature anomalies, ten-year low-pass filtered data by means of a multivariate statistical regression model using simultaneously volcanic (SVI* or DVI), solar (sunspot relative numbers, SRN, or solar oscillation hypothesis, TGI) and CO_2 forcing.

Note: For details see Schönwiese (1987); Schönwiese and Malcher (1987)

Acknowledgement

This study was supported by the German (FRG) Climate Research Programme (BMFT, project numbers KF 1009 0 and 2012 8).

References

Angell, J.K., Korshover, J. 1984. 'Seasonal tropospheric and stratospheric temperature anomalies for the northern hemisphere, 1958–1984', *CDIC Numeric Data Collection NDP-008*, National Laboratory, US Department of Energy, Oak Ridge.

Angell, J.K., Korshover, J. 1985. 'Surface temperature changes following the six major volcanic episodes between 1780 and 1980', *J. Clim. Appl. Met.*, 24, 937–51.

Bryson, R.A., Goodman, B.M. 1980. 'Volcanic activity and climatic changes', *Science* 207, 1041–44.

Gilliland, R.L. 1982. 'Solar, volcanic, and CO_2 forcing of recent climatic changes', *Climatic Change* 4, 111–31.

Groveman, B.S., Landsberg, H.E. 1979. 'Reconstruction of the northern hemisphere temperature: 1579–1880', Publ. 79 181/182, Dep. Meteorol. Univ., Maryland.

Hammer, C.U., Clausen, H.B., Dansgaard, W. 1981. 'Past volcanism and climate revealed by Greenland ice cores', *J. Vulcanol. Geoth. Res.* 11, 3–10.

Hansen, J., Johnson, D., Lacis, A., Lebedeff, S., Lee, P., Rind, D., Russel, G. 1981.

'Climate impact of increasing atmospheric carbon dioxide', *Science* 213, 957–66.

Jäger, H., Reiter, R., Carnuth, W., Funk, W. 1984. 'El Chichón cloud over Europe', *Geof. Int.* 23, 243–57.

Jones, P.D. 1985a. 'Northern hemisphere temperatures 1851–1984', *Climate Monitor* 14, 14–21 (updatings, also mean Arctic and central England data, same journal).

Jones, P.D. 1985b. 'Arctic temperatures 1851–1984', *Climate Monitor* 14, 43–50.

Labitzke, K., Brasseur, G., Naujokat, B., De Rudder, A. 1986. 'Long-term temperature trends in the stratosphere. Possible influence of anthropogenic gases', *Geophys. Res. Letters* 13, 52–5.

Lamb, H.H. 1970. 'Volcanic dust in the atmosphere; with a chronology and assessment of its meteorological significance', *Phil. Trans. R. Met. Soc.* A266, 425–533.

Lamb, H.H. 1983. 'Update of the chronology of assessments of the volcanic dust veil index', *Climate Monitor* 12, 79–90.

McCormick, M.P., Swissler, T.J., Chu, W.P., Fuller Jr., W.H. 1978. 'Post-volcanic aerosol decay as measured by lidar', *J. Atmos. Sci.* 35, 1296–1307.

Miles, M.K. Gildersleeves, P.B. 1978. 'Volcanic dust and changes in northern hemisphere temperature', *Nature* 271, 735–6.

Newhall, C.G., Self, S. 1982. 'The volcanic explosivity index (VEI): an estimate of explosive magnitude for historical volcanism', *J. Geophys. Res.* 87, 1231–8.

Pollack, J.B., Toon, O.B., Summers, A., Baldwin, B., von Camp, W. 1976. 'Volcanic explosions and climatic change: a theoretical assessment', *J. Geophys. Res.* 81, 1071–83.

Rampino, M.R., Self, S., 1984. 'Sulfur-rich eruptions and stratospheric aerosols', *Nature* 310, 677–9.

Reiter, R., Jäger, H. 1986. 'Results of 8-year continuous measurements of aerosol profiles in the stratosphere with discussion of the importance of stratospheric aerosols to an estimate of effects on the global climate', *Meteorol. Atmos. Phys.* 35, 19–48.

Schönwiese, C.D. 1983. 'Northern hemisphere temperature statistics and forcing. Part A: 1881–1980 AD', *Arch. Met. Geoph. Biocl.* Ser. B, 32, 337–60.

Schönwiese, C.D. 1984. 'Northern hemisphere temperature statistics and forcing. Part B: 1579–1980 AD', *Arch. Met. Geoph. Biocl.* Ser. B, 35, 155–78.

Schönwiese, C.D. 1986. 'Zur Parameterisierung der nordhemisphärischen Vulkantätigkeit seit 1500', *Meteorol. Rdsch.* 39, 126–32.

Schönwiese, C.D. 1987. 'Observational assessments of the hemispheric and global climate response to increasing greenhouse gases', *Contr. Phys. Atmosph.* 60, 48–64.

Schönwiese, C.D., Malcher, J. 1987. 'The CO_2 temperature response. A comparison of the results from general circulation models with statistical assessments', *J. Climatology* 7, 215–29.

Sear, C.B., Kelly, P.M. 1982. 'The climatic significance of El Chichón, *Climate Monitor* 11, 134–9.

Simkin, T., Siebert, L., McClelland, L., Bridge, D., Newhall, C., Latter, J.H. 1981 *Volcanoes of the world* Smithsonian Institution, Hutchinson, Stroudsbourg.

Toon, O.B., Pollack, J.B. 1980. 'Atmospheric aerosols and climate', *Amer. Scientist* 68, 268–78.

Chapter 3

Large-scale precipitation fluctuations: a comparison of grid-based and areal precipitation estimates

P.D. Jones   University of East Anglia, United Kingdom

Introduction

Most studies of the large-scale changes in climate that have taken place over the period of instrumental records are made using surface air temperature and sea-level pressure data. Despite the fact that precipitation changes have a far greater effect on man through effects on agriculture, hydrology and water resources, studies of precipitation fluctuations on continental and hemispheric scales are rare.

Possible or projected changes in precipitation in the future due to increases in atmospheric CO_2 will also have a greater potential impact than temperature changes. The rationale behind studying large-scale precipitation changes lies with the estimates of future precipitation levels being made in the perturbed runs (e.g., with increased levels of atmospheric CO_2) of general circulation models (GCMs). GCMs are considered more reliable when the results are averaged over the largest spatial scales. The spatial and temporal variability of measured precipitation need to be known in order to assess the significance of any implied changes. It is important, therefore, that we begin to study the historical record of precipitation variability in order to place possible or projected future conditions into a longer-term context.

Data and method

In related work on large-scale temperature averages, Jones *et al.* (1986a; 1986b) have interpolated station temperature values onto a regular latitude–longitude grid. Gridding the data prevents regions with high station density unduly influencing regional scale temperature averages. It is not possible to interpolate using raw temperatures because neighbouring stations may be at different elevations and may calculate monthly mean temperature using different formulae. Both these problems are easily overcome by reducing all the temperature data to anomalies from a common reference period. Jones *et al.* (1986a; 1986b) used 1951–70 as the common period for both the northern and southern hemispheres.

Station precipitation data, together with sea-level pressure data, form the only other large data sets comparable with station temperature data. Large-scale precipitation averages may be used as a measure of the hemispheric hydrological cycle and might reflect some of the effects of possible CO_2-induced climatic change.

Precipitation data, however, are much more difficult to deal with than temperature data. Precipitation data are often positively skewed (Fiering and Kuczera 1982) and this can influence analyses if it is not adequately taken into account. In order to achieve some form of interpolation onto a regular grid, it is necessary to transform the data. One method is to use percentage departures from a reference grid (Corona 1978; 1979; Gruza and Apasova 1981) but this poses problems when the data are not normally distributed.

Bradley *et al.* (1987) accommodated this problem by converting data from each station to probability estimates using the gamma distribution, in a manner described by Thom (1958; 1966). Previous work has demonstrated that the gamma distribution provides a good fit to precipitation data and enables precipitation amounts to be quite accurately expressed as probabilities at each station (Ropelewski *et al.* 1985). The gamma distribution is a two-parameter probability distribution:

$$f(x) \, \frac{1}{\beta^{\gamma} \, \lceil \gamma} \, x^{\gamma - 1} e^{-x/\beta} \quad \beta > 0, \ \gamma > 0 \tag{1}$$

where x is the monthly, seasonal or annual precipitation, γ is the 'shape' parameter, β is the 'scale' parameter and $\lceil$ is the gamma function. The probability is obtained from the cumulative density function,

$$F(x) = \int_{0}^{x} f(t)dt \tag{2}$$

calculation of which necessitates computation of the incomplete gamma integral. Simple algorithms are available for this purpose (e.g., Lau 1980).

Shape and scale parameters were calculated, using maximum likelihood estimates (Thom 1958), for monthly, standard seasons and annual precipitation at 1487 stations using data for the reference period 1921–60 (selected as the period that would retain the maximum number of stations). Examples of the shape and scale parameters for four representative stations are given in Table 3.1. (Note that the mean precipitation is equal to $\gamma \beta$.)

Interpolation of the station probability values was accomplished by using a negative exponential function;

$$W = e^{-d2/4c} \tag{3}$$

where d is the distance from the station to grid point and c (the weight constant) was set to 500. If no station were located within 500 km of a grid point no value was computed. One station value became the grid point value if this was all that was available. In order that stations in data-sparse regions did not have undue influence on large-scale averages, a single station could not influence more than the four nearest grid points. A maximum of 1410 grid points (45.8 per cent of the equal area grid used by Bradley *et al.* 1987) were able to have interpolated values out of the maximum possible of 3077 for the northern hemisphere.

Figure 3.1 shows the time series of the mean probability values over the northern hemisphere. The degree of coverage changes through time and at least half of the maximum number of contributing gridpoints is not reached until 1883 (Bradley *et al.* 1987). The series enables trends and extremes to be analysed but in this form the results cannot be compared with model results. It is necessary therefore to convert the data back to millimetres from probability estimates.

Table 3.1 Shape (γ) and scale (β) parameters for four representative stations

Station Country	Inverness UK		Cherrapunji India		Kayes Mali		Avalon Pleasure Pier, USA	
	γ	β	γ	β	γ	β	γ	β
J	6.4	8.1	0.5	52.5	0.4	1.8	1.0	57.4
F	5.3	7.2	0.8	50.9	0.4	1.4	1.0	78.8
M	4.4	7.7	1.1	157.1	1.1	0.2	0.7	65.7
A	3.5	12.1	3.5	187.4	0.3	23.4	0.8	34.4
M	5.0	10.1	4.1	389.8	0.4	45.5	0.4	17.6
J	3.4	14.6	10.1	285.5	5.6	16.9	0.5	2.9
J	4.0	18.4	7.0	347.6	8.3	19.9	2.5	0.1
A	3.0	26.1	7.7	226.8	6.3	39.1	0.5	1.9
S	4.2	14.2	5.2	232.3	5.7	29.4	0.3	8.8
O	4.0	17.1	1.4	328.5	1.2	35.3	0.4	38.5
N	3.0	18.0	0.3	160.2	0.3	10.1	0.5	52.5
D	4.2	13.1	0.3	37.9	0.8	0.4	0.7	85.0
MAM	13.4	9.4	8.7	278.7	0.4	61.5	1.8	45.2
JJA	9.5	21.3	25.1	280.9	13.3	37.8	0.5	4.9
SON	12.1	15.1	6.9	249.0	5.8	37.3	0.8	53.5
DJF	13.2	11.0	1.1	67.6	0.4	3.2	3.3	59.3
Annual	64.6	10.1	43.0	262.1	20.3	36.9	4.9	65.6

Notes: MAM = March, April, May total etc.
Mean rainfall = $\gamma \beta$

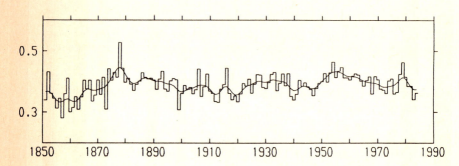

Figure 3.1 Annual precipitation index (mean value of the gamma probability distribution at all available grid points) for nothern hemisphere land masses

Note: The smooth line on this and subsequent plots is designed to show decadal and longer time-scale changes.

Transformation of the probability data back to millimetre units is not a trivial task for two reasons. First, shape and scale parameters must be estimated for each grid point and second an efficient scheme for the inversion of equation (2) is required. In this chapter, we present one method of retransforming the gridded precipitation data to millimetres (mm).

Shape and scale parameters for each grid point (γ_g, β_g) were interpolated from the station values using simple weighting schemes:

$$\gamma_g = \sum_{i=1}^{N} \gamma \frac{i}{N} \tag{4}$$

$$\beta_g = \frac{\sum\limits_{i=1}^{N} \dfrac{\beta_i}{d_i}}{\sum\limits_{i=1}^{N} \dfrac{1}{d_i}} \tag{5}$$

where γ_i and β_i are the shape and scale parameters for the N stations within 500 km of a grid point. The d_i's distances between the stations and the grid point. The form of the interpolation was made after subjectively considering the spatial variation of variations of γ and β. Neighbouring station values in neighbouring precipitation sites, which may be the result of differing station evaluations, are reflected by changes in the scale parameter. A recently developed algorithm (see also Best and Roberts 1975), accurate to 0.1 mm, was used to transform the probabilities back to millimetres.

Results

Comparison with large-scale precipitation series

For a number of regions of the world areal precipitation series have been developed from station data. The gridded data will be compared with two series from contrastingly-sized regions, that for England and Wales (Wigley *et al.* 1984, revised and updated in Wigley and Jones 1987) and that for the contiguous United States (Diaz and Quayle 1980, updated by National Climate Data Center, Asheville, NC). For the England and Wales region four grid points were incorporated into the average.

Figure 3.2 shows the annual precipitation totals for England and Wales calculated by three different methods; (a) areal series (Wigley *et al.* 1984), (b) gridded data in millimetres and (c) gridded data in probability units. Figure 3.3 shows contiguous United States rainfall calculated by the same three methods. For this region both the gridded data sets extend back to 1851 while the areal series (Diaz and Quayle 1980) only begins in 1895.

Correlation coefficients between the three different methods of calculation for annual and seasonal (not shown) precipitation are given in Table 3.2. For both regions, the correlations between the areal series and the gridded millimetre data are higher than with the gridded probability data, although only in the June to August season for the US region is there a significantly higher correlation. Correlations tend to be lower in the summer half of the year, a fact which is probably related to the greater amount of convective precipitation in this season. A greater raingauge density is required with this type of precipitation in order to achieve a specified accuracy in areal precipitation estimation. Considerably fewer precipitation records were used in the calculation of the gridded data than for both areal series.

One of the major problems with the gridded data is evident in the United States

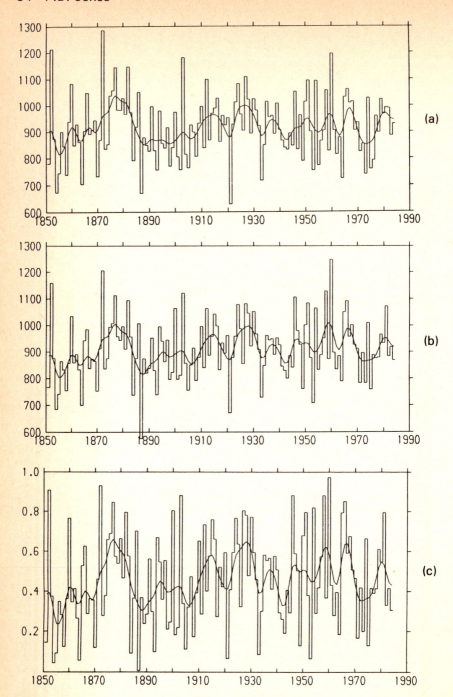

Figure 3.2 Mean annual precipitation for England and Wales calculated by three different methods: (a) Wigley *et al.* (1984); (b) gridded data in millimetres; (c) gridded data in gamma probability units.

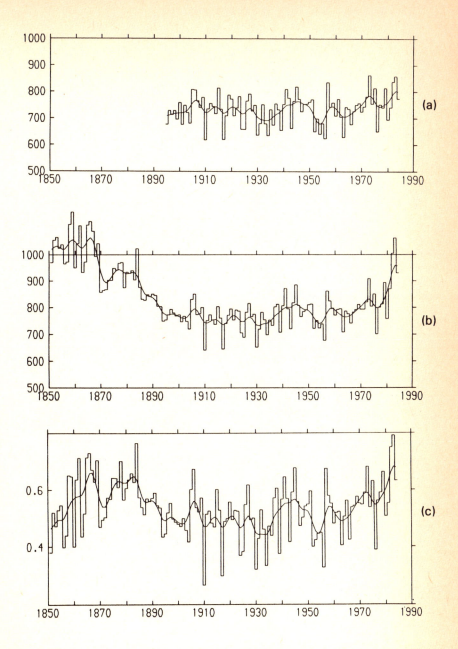

Figure 3.3 Mean annual precipitation for the continuous United States calculated by three different methods (a) Diaz and Quayle (1980); (b) gridded data in millimetres; (c) gridded data in gamma probability units.

Table 3.2   Correlations between seasonal and annual estimates of regional precipitation

Season	a) England and Wales, 1851–1980			b) Contiguous United States, 1895–1980		
	A v B	A v C	B v C	D v E	D v F	E v F
MAM	0.935	0.911	0.980	0.944	0.926	0.977
JJA	0.954	0.938	0.982	0.903	0.746	0.829
SON	0.954	0.944	0.985	0.931	0.929	0.960
DJF	0.949	0.932	0.986	0.893	0.871	0.950
Annual	0.941	0.932	0.986	0.911	0.914	0.967

A = Wigley *et al.* (1984)
B = Gridded data set in millimetres
C = Gridded data set in probability units
D = Diaz and Quayle (1980)
E = Gridded data set in millimetres
F = Gridded data set in probability units

data in Figure 3.3b. Prior to 1895 and for years since 1980 annual precipitation values were relatively high. This is not a real feature but is related to a reduction in the number of grid points in the precipitation data set where interpolation could be made. Between 1895 and 1980 a constant 91 grid points can be interpreted. The gridpoints which are not available before 1895 are in the drier parts of the country between the Rocky Mountains and the Mississippi River, a fact which can be easily confirmed by mapping the available station data for years before 1890 (Bradley *et al.* 1985).

The problem is not apparent with the probability version of the gridded data set because here each interpolated grid point is scaled to equivalent probability units. Averaging over the entire country assumes that all unrepresented regions of the country, before 1895, have the mean of all the represented regions. This reduces the effect of variations in data availability for regions with different mean precipitation. A similar assumption is made in the calculation of hemispheric temperature anomalies in the series produced by Jones *et al.* (1986a; 1986b) and in some regional precipitation series (Lamb 1982). For temperature the grid is scaled to equivalent units by expressing the temperature as an anomaly value.

Northern hemisphere mean precipitation

Figure 3.1 shows the mean precipitation over the northern hemisphere expressed in probability units. If we attempt to calculate a similar series in millimetre units the problem of changing coverage before 1921 and after 1960 will affect the series as in the United States example in Figure 3.3b.

One way of overcoming the problem is to express each gridpoint as a percentage of the 1921–60 reference period then average all the gridpoint percentages in a particular season or year. This mean percentage may then be expressed in millimetres by reference to the hemispheric mean value of the reference period. Mathematically this can be written:

$$p_j = \bar{p}_j \sum_{i=1}^{N} \frac{g_{ij}}{p_i} \qquad j = 1, 4 \text{ seasons} \tag{6}$$

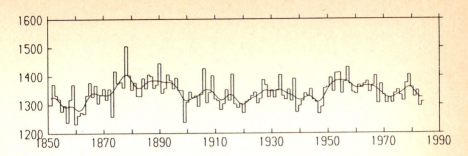

Figure 3.4 Annual precipitation for the northern hemisphere landmasses calculated in millimetres

Notes: For details of the calculation procedure see text.

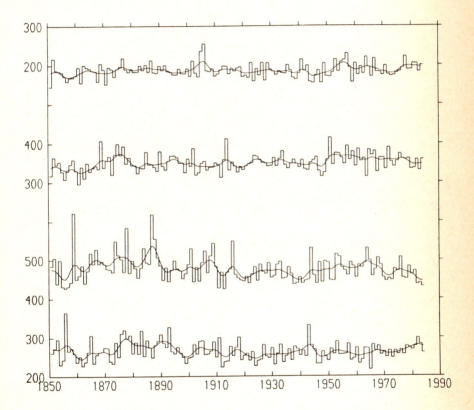

Figure 3.5 Seasonal precipitation for the northern hemisphere landmasses calculated in millimetres

Note: For details of the calculation procedure see text. Seasons are, from bottom to top, MAM, JJA, SON, DJF.

Table 3.3   Correlations between millimetre and probability estimates of the northern hemisphere mean precipitation, 1851–1980

Season	Correlation
MAM	0.740
JJA	0.684
SON	0.819
DJF	0.743
Annual	0.898

Table 3.4   Comparison of estimates of the average precipitation (mm) over the northern hemisphere

Season	This Work (1921–60)	Jaeger NH (land, 1931–60)	Jaeger NH (land and sea, 1931–60)
MAM	259	136	210
JJA	469	265	302
SON	354	178	264
DJF	184	96	202
Annual	1343	756	978

where $\bar{p}_j$ is the seasonal or annual mean precipitation over the northern hemisphere, $\bar{p}_i$ are the seasonal or annual mean precipitation at each of N grid points and g_{ij} the gridded precipitation values in millimetres.

The annual hemispheric mean series calculated in this way is plotted in Figure 3.4 and the seasonal series are plotted in Figure 3.5. There are slight differences between these series and those calculated with the probability data. Correlation coefficients between the two series are given in Table 3.3. None of the trends evident in the probability series discussed by Bradley *et al.* (1987) is altered by expressing the series in millimetres. Differences in some of the extreme years are noticeable. For the annual data, 1979 was the 'wettest' year in the probability series but with the millimetre version nine separate years between 1878 and 1979 are wetter than 1979.

The 1921–60 reference period means precipitation for the northern hemisphere calculated using the gridded millimetre data can be compared with estimates made by Jaeger (1976). Jaeger's estimates were made using data from climatic atlases and tables of normals relating to the 1931–60 reference period. Seasonal and annual totals are compared in Table 3.4. Note that, because of the non-linear nature of the gamma distribution which has been applied to the original station precipitation data, the sum of the four seasons is not constrained to be equal to the annual total. The present gridded data set estimates the mean northern hemisphere precipitation for the land areas to be just under twice that of Jaeger. Overestimation is to be expected because regions not represented in this analysis are mainly arid areas. The major omitted regions are Arctic areas of North America, Greenland and the Soviet Union, the central Asian region of western China, Mongolia and Afghanistan, the Middle East and parts of central Northern Africa.

For most of the northern hemisphere, however, the data set provides a measure of large-scale and regional-scale spatial precipitation distribution and, more importantly, precipitation variability with which to compare model estimates.

Conclusions

The gridded data have been shown to produce reliable estimates of areal precipitation for two regions of contrasting size, England and Wales and the contiguous United States. Provided the region under study does not experience marked changes in spatial coverage through time reliable estimates of areal precipitation can be obtained from the gridded data.

Initial attempts have been made to estimate hemispheric precipitation trends, although these are hampered to some degree by the varying availability of gridpoint data through time. Estimates of the reference period mean precipitation over land areas of the northern hemisphere are compared with previous work. Although the lack of gridpoint data in many arid areas implies that the data set overestimates mean hemispheric precipitation, the data set should still prove useful for the spatial validation of GCM control runs.

The advantage of the gridded data over the Jaeger estimates is that apart from studying mean precipitation over the 1921–60 period, the variability within months, seasons and years may also be studied. No real world data set has been available for comparison with GCM precipitation variability over the continents on a regional and hemispheric scale.

Acknowledgements

The work described here was supported by the Carbon Dioxide Division, US Department of Energy, through grant DE-FG02-85ER 60316. The author thanks colleagues at the Climatic Research Unit for helpful comments on the manuscript.

References

Best, D.J., Roberts, D.E. 1975. 'The percentage points of the chi-squared distribution', *Appl. Statist.* 24, 385–8.

Bradley, R.S., Kelly, P.M., Jones, P.D., Goodess, C.M., Diaz, H.F. 1985. *A Climatic Data Bank for Northern Hemisphere Land Areas, 1851–1980*. US Dept of Energy, Carbon Dioxide Research Division, Technical Report TRO17.

Bradley, R.S., Diaz, H.F., Eischeid, J.K., Jones, P.D., Kelly, P.M., Goodess, C.M. 1987. 'Precipitation fluctuations, over Northern Hemisphere land areas since the mid-19th century', *Science* 237, 171–5.

Corona, T.J. 1978. 'The interannual variability of Northern Hemisphere Precipitation', Environmental Research Paper no. 16, Colorado State University, Fort Collins, CO.

Corona, T.J. 1979. 'Further investigations of the interannual variability of Northern Hemisphere continental precipitation', Environmental Research Paper no. 20, Colorado State University, Fort Collins, CO.

Diaz, H.F., Quayle, R.G. 1980. 'The climate of the United States since 1895: spatial and temporal changes', *Mon. Wea. Rev.* 108, 249–66.

Fiering, M.B. Kuczera, G. 1982. 'Robust estimators in hydrology' in *Scientific Basis of Water-Resources Management*, Geophysical Study Committee, National Research Council, National Academy of Sciences, Washington, DC USA, 85–94.

Gruza, G.V., Apasova, Y.G. 1981. 'Climatic variability of the Northern Hemisphere precipitation amounts', *Meteorologiya i Gidrologiya* 5, 5–16.

Jaeger, L. 1976. 'Monatskarten des Niederschlags für die ganze Erde', *Berichte des deutschen Wetterdienst* 139, Offenbach,

Jones, P.D., Raper, S.C.B., Bradley, R.S., Diaz, H.F., Kelly, P.M., Wigley, T.M.L. 1986a. 'Northern Hemisphere surface air temperature variations, 1851–1984', *Clim. Appl. Met.* 25, 161–79.

Jones, P.D., Raper, S.C.B., Wigley, T.M.L. 1986b. 'Southern Hemisphere surface air temperature variations, 1851–1984', *J. Clim. Appl. Met.* 25, 1213–30.

Lamb, P.J. 1982. 'Persistence of subsaharan drought', *Nature* 299, 46–9.

Lau, C. 1980. 'A simple series for the Incomplete Gamma Integral', *Appl. Statist.* 29, 113–14.

Ropelewski, C.F., Janowiak, J.F., Halpert, M.S. 1985. 'The analysis and display of real time surface climate data', *Mon. Wea. Rev.* 113, 1101–6.

Thom, H.C.S. 1958. 'A note on the gamma distribution', *Mon. Wea. Rev.* 86, 117–22.

Thom, H.C.S. 1966. 'Some methods of climatological analysis', *World Meteorological Organization Technical Note no. 81*, WMO no. 199 TP. 103, Geneva, Switzerland.

Wigley, T.M.L., Lough, J.M, and Jones, P.D. 1984. 'Spatial patterns of precipitation in England and Wales and a revised, homogeneous England and Wales precipitation series', *J. Climatol.* 4, 1–25.

Wigley, T.M.L., Jones, P.D. 1987. 'England and Wales precipitation: A discussion of recent changes in variability and an update to 1985', *J. Climatol.* 7, 231–46.

Chapter 4

The Meteorological Office Historical Sea Surface Temperature data set

D.E. Parker and C.K. Folland   Meteorological Office, United Kingdom

Introduction

The Meteorological Office historical sea surface temperate data set was created to fulfil diverse needs in research into practical long-range forecasting and into climatic fluctuations observed since the mid-nineteenth century. This chapter provides a brief account of the origins, construction and quality control of the data set. For greater details the reader is referred to the new sea surface temperature atlas being produced by the Meteorological Office jointly for the US Navy with the Massachusetts Institute of Technology (MIT) (Bottomley *et al.* 1988).

The value of the data set has been substantially increased by the recent application of geographically and seasonally varying instrumental corrections to all data up to 1941 to compensate for biases affecting uninsulated canvas bucket data which are thought to have been predominant until then. The corrections were computed using a model of the physics of these buckets and their environment. Here we describe the principles underlying the use of the model and illustrate some of the apparently beneficial results of applying the corrections derived from it.

Origins

Ships' observations have been systematically recorded since the mid-nineteenth century following the plan of Maury and Glaisher (1852) to cover the whole world with observing stations. Many of these records were stored in archives of ships' logbooks until the 1960s when they were keyed into computerised datasets such as the 'TDF-11' data set created at the United States Climatological Center, Asheville, (NC).

These data sets were supplemented by records for more recent years which have been exchanged internationally under the World Meteorological Organization's 'Resolution 35' which was adopted in 1963. The Meteorological Office main marine data bank (MOMMDB) (Shearman 1983) was formed by combining these data sources, with careful checks being made to remove any duplicate observations. The Meteorological Office historical sea surface temperature (MOHSST) data set has been created by extracting the sea surface temperature observations from the MOMMDB. Similar historical data sets have also been created for nighttime (MOHMATN) and daytime marine air temperature. None of these data sets yet includes satellite data.

Construction and coverage

The MOHSST data set contains monthly values for 5° latitude × 5° longitude areas, beginning at January 1854 and continuing up to the present. The basic processes involved in the construction of the data set are outlined in Figure 4.1, which includes related activities such as creation of climatologies and derivation of instrumental and procedural corrections. Corrections are not applied to the data set directly, but are stored separately, and used as necessary.

A failure many years ago to receive certain magnetic tapes resulted in a sparsity of data in the main marine data bank for large areas of the Pacific in 1961–72. The gaps were filled in MOHSST by using analysed 5° latitude × 5° longitude data received from MIT, which had been derived from the consolidated data set assembled by the US Navy Fleet Numerical Oceanography Center. The missing observations have recently been incorporated into MOMMDB and will be included in the next update of MOHSST.

Figure 4.2a–c give a measure of the coverage of the MOHSST data set in 1861–70, 1911–20 and 1971–80, respectively. Geographical coverage is very good in the Atlantic and Indian Oceans for the twentieth-century and fair there for the nineteenth-century portion of the data set, but very poor in the Pacific for before 1900 and in the Southern Ocean south of 40°S throughout the period of the data

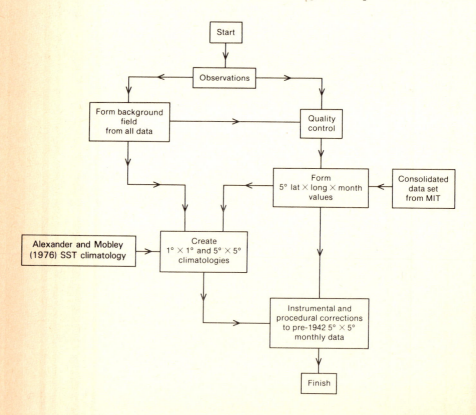

Figure 4.1   Sea surface temperatures: overall processing

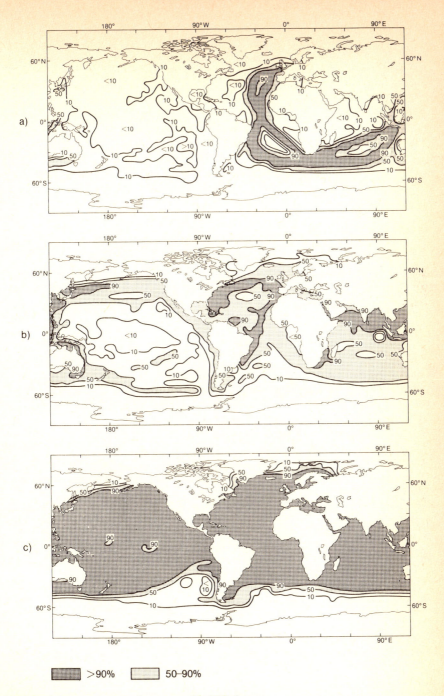

Figure 4.2 Percentage of seasons with SST data
(a) 1861–70
(b) 1911–20
(c) 1971–80

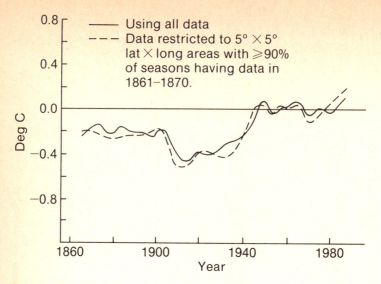

Figure 4.3 Global SST w.r.t. 1951–80. Values plotted at end date of 10.25-year triangular filter.

Note: Bucket corrections applied to SST up to 1941.

set. Nevertheless Figure 4.3 shows that the limitations of coverage have not greatly affected estimates of longest-term changes of global mean sea surface temperature. Restriction of the data to the heavily hatched areas in Figure 4.2a (90 per cent or better coverage in 1861–70) only caused the estimate of the global trends to be changed from the solid curved to the dashed curve in Figure 4.3. The reason for this insensitivity to coverage is that the long-term trends appear to have occurred in unison over virtually the entire globe (Bottomley *et al.* 1988). Estimates of hemispheric long-term trends are almost as insensitive to the changes of coverage as are the estimates of global trends.

The need for a globally complete climatology (for, for example, numerical modelling purposes) was, however, met by merging a climatology published by Alexander and Mobley (1976) which itself was only complete because the authors essentially interpolated in the Southern Ocean between the southern limit of observed data and an assumed ice-edge temperature of −1.7°C.

Quality control

The quality-control procedures were designed to treat individual poor data and to compensate as well as possible for irregular distribution of data in space and time. First, observations of sea surface temperature outside the range −2 °C to +37 °C were rejected. Rejection of all values below −2 °C will have caused estimates of true sea surface temperature near this limit to have been slightly positively biased, by including only observations with positive errors. The remaining sea surface temperatures were then converted to differences from a background field of approximate climatological averages. This field has been formed, on 1° latitude ×

Table 4.1 Definition of 'months' used in processing SST and NMAT data from the MOMMDB

Month	Pentads	Calendar Equivalent
January	1–6	1–30 January
February	7–12	31 January – 1 March (in all years)
March	13–18	2–31 March
April	19–24	1–30 April
May	25–30	1–30 May
June	31–36	31 May–29 June
July	37–42	30 June–29 July
August	43–49	30 July–2 September
September	50–55	3 September–2 October
October	56–61	3 October–1 November
November	62–67	2 November–1 December
December	68–73	2–31 December

1° longitude × pentad resolution, by averaging all data, provided they lay between −2 °C and +37 °C, for all years (1854–1981) in the appropriate box and pentad, and applying a complex timewise and spacewise smoothing procedure (Bottomley *et al.* 1988). Individual sea surface temperatures were rejected if they differed by more than ± 6 °C from the background field. Tests with broader limits (e.g., ± 7 °C) showed little difference in the final results even in the eastern tropical Pacific where El Niño events are sometimes associated with marked positive anomalies. The remaining sea surface temperatures were averaged separately over each available 1° latitude × 1° longitude box for each available pentad (e.g. 1–5 January 1927) and the results were converted to differences of 'anomalies' from the background field.

Then, for a given 5° latitude × 5° longitude area and month, the 1° latitude x 1° longitude pentad anomalies (a maximum of 150 or 175 values) were censored using 'Winsorisation' (Afifi and Azen 1979): i.e. before the anomalies were averaged, all values in the outer quarters of the distribution were replaced by the value of the relevant outer quartile of the distribution. This tempered the effects of the outliers while retaining some information about the skewness of the distribution. If fewer than 16 anomalies were used, these were subjected to consistency tests, and flags were set if the tests were failed. If there were fewer than four 1° latitude × 1° longitude pentad anomalies in a month, Winsorisation could not be carried out, and a simple average was calculated and flagged. The 5° latitude × 5° longitude average anomaly so calculated was added to the 5° latitude × 5° longitude average of the background field to give the average sea surface temperature for that area and month.

Note that working on 1° latitude × 1° longitude × pentad resolution can avoid considerable local errors in the 5° area anomalies in the presence of irregular distribution of data in space and time. If, for example, an observation at 41 °N 18 °W on 31 May 1856 were the only one in May 1856 in the area 40°–45 °N 15°–20 °W, it would almost certainly be warmer than the average for that area and month, even if it were colder than the local average for 31 May. The procedure also avoids slowly varying biases which would otherwise result if shipping routes changed systematically from one part of a 5° area to another part. Note also that the 'months' used in the data set are pseudo-months as in Table 4.1.

Instrumental corrections

Systematic instrumental biases are not treated by the above quality-control procedures. The instrumental corrections needed are of the order of several tenths of a degree and relate to the apparently sudden change-over from uninsulated canvas buckets (with a few engine intake data) before the Second World War to far more engine intake data during the War. Since then, engine intake data appear to have been dominant, or at least the use of uninsulated buckets has not been widespread (although some British ships did use them). Nowadays most of the bucket data are from insulated buckets, as is evidenced by a comparison between bucket and other data for 1975–81 which yielded a global annual averaged difference of only 0.08 °C (Folland, Parker and Newman 1985), with the bucket data reading colder, though the differences tended to be greater in winter (up to 0.2 °C) and reversed in summer.

Corrections to compensate for the use of uninsulated buckets were calculated using a model of the physics of an uninsulated canvas bucket. The model takes into account the heat fluxes arising from the following causes:

1. the difference between the external air temperature and the temperature of the water in the bucket;
2. the difference between the deck temperature and the temperature of the water in the bucket, affecting conduction through the base of the bucket;
3. the difference between the atmospheric vapour pressure and the saturation vapour pressure at the temperature of the water in the bucket;
4. the strength of the wind;
5. the radiation incident on the bucket.

The combination of (3) and (4) in particular renders mid-latitude winter uninsulated bucket SST too cold; whereas (5) and to a lesser extent (2) can make mid-latitude summer uninsulated bucket SST less cold or even a little too warm, resulting in spurious annual cycles of pre-war SST anomalies relative to postwar SST climatology (Figure 4.4).

Application of the model was beset by a number of uncertainties, which essentially related to the actual environmental conditions affecting the bucket; the thermal capacity and other physical properties of the water-filled bucket; and the time which elapsed between the removal of the bucket from the sea and the reading of the thermometer, which we shall call the 'duration of observation'. The principle used to offset these uncertainties was that the corrections derived from the model should compensate for the seasonal variations of heat flux, and therefore minimise the spurious annual cycles in SST (Figure 4.4) in a wide variety of regions with differing annual courses of climate. The environmental conditions were estimated from 1951–80 monthly climatologies of marine temperatures based on MOHSST and MOHMATN, and 1949–79 monthly climatologies of humidity, wind and insulation obtained from MIT, but the winds, air temperatures and radiation were first adjusted for the influence of the ship both during hauling (assumed to last one minute) and subsequently on deck. The greatest uncertainty attaches to the duration of observation (Brooks 1926), though the consensus of instructions to observers would indicate three to five minutes including hauling (e.g. Meteorological Office 1881; page 1904). The duration of observation was not, therefore, chosen a priori: instead, the model was used to compute the corrections, and the duration of observation for which the corrections minimised the spurious annual cycles for

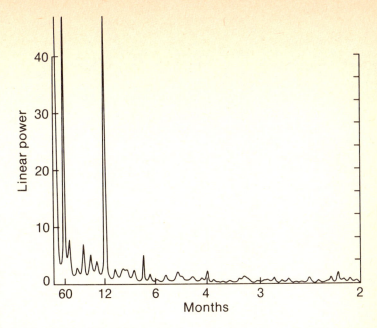

Figure 4.4 Power spectrum of uncorrected SST anomalies, 1901–41, North Atlantic (the y-axis is in values proportional to $(°C)^2$)

1881–1941 as a whole and for constituent 15-year periods was noted. The sensitivity of the calculations to a variety of uncertainties was then tested. For example, effective wind strengths of 100 per cent, one-third or 15 per cent of climatology while on deck were tried, as were full climatological insolation, 50 per cent shading from direct insolation, or total shading from direct insolation. The model was found to produce corrections which were rather insensitive to these assumptions if the model was integrated until the spurious annual cycles were minimised. When this was done it was found that the duration of observation required to minimise the spurious annual cycles remained between three and five minutes in most regions; and the resulting global and hemispheric annual average corrections, which are important for determination of long-term climatic changes, did not vary by more than about 0.1 °C. However, corrections derived in a similar manner for before 1875 do not have these characteristics and may therefore be much less reliable. Also when the model was applied for 1881–1941 to a small area in the Gulf Stream, where air–sea temperature differences are large and variable, the duration of observation required was seven minutes. A better procedure in future could be to use available data on environmental conditions at the time of individual SST observations to derive the corrections, though climatological estimates of, e.g., radiation would still be needed, and the non-linear relationship between environmental conditions and the corrections is still likely to make the corrections less reliable in areas such as the Gulf Stream.

Figure 4.5 shows the corrections for June and December based on a 'standard' version of the model considered to be most realistic on observational and theoretical grounds, i.e. with wind strength on deck equal to one-third of climatology, and

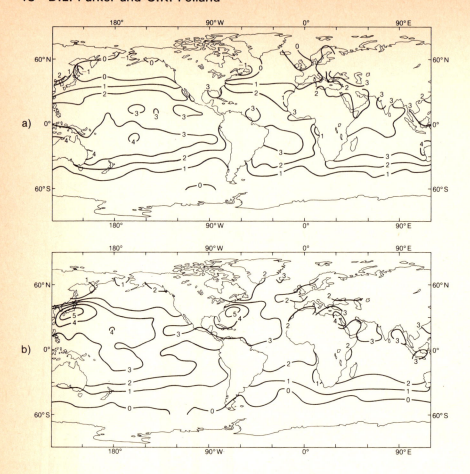

Figure 4.5 Corrections to uninsulated bucket SST (tenths °C):
(a) June
(b) December

with 50 per cent shading from direct insolation. The duration of observation was assumed to be four minutes to minimise most of the spurious annual cycles for 1881–1941. As expected, the corrections are greatest in winter, especially in areas where the sea is much warmer than the air (e.g., over the Gulf Stream and Kuroshio). In summer they can be negative, indicating that warming of the bucket by insolation slightly exceeds cooling from evaporation and conduction, especially when air–sea temperature differences are also very small.

The effect of the removal of the spurious annual cycles is illustrated in Figure 4.6 which shows the extent to which the 'standard' corrections have increased the argument between climatic trends for the separate seasons for the northern hemisphere. The curves are plotted against the end-date of a ten-year filter: it is clear, therefore, that there is even some value in the corrections before about 1880, though the reduction in the spurious annual cycles then is too small.

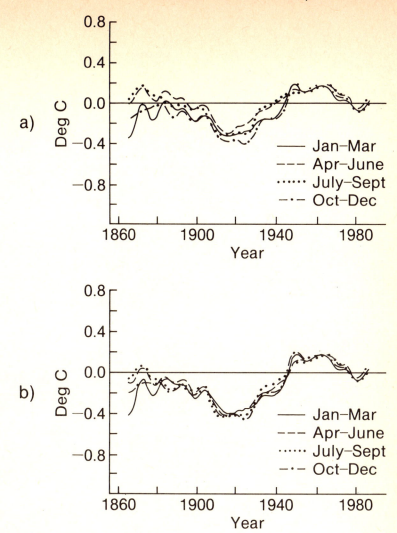

Figure 4.6 Northern hemisphere SST
(a) 0.3 °C added to data up to 1941;
(b) Bucket-model corrections up to 1941.

Conclusion

The current version of the Meteorological Office historical sea surface temperature data set (MOHSST3) has been described. Its two most important features are the analyses of basic data in terms of anomalies on a 1° latitude × 1° longitude space scale to remove biases due to changing ships' tracks, and the physically based corrections applied to data for before the early 1940s. The latter are now being refined to include the non-linear effects of the climatological distribution of wind speed in each region and calendar month.

References

Afifi, A.A., Azen, S.P. 1979. *Statistical Analysis: A Computer Oriented Approach*, 2nd edn, New York, Academic Press.

Alexander, R.C., Mobley, R.L. 1976. 'Monthly average sea-surface temperatures and ice-pack limits on a 1° global grid', *Mon. Wea. Rev.* 104, 143–8.

Bottomley, M., Folland, C.K., Hsiung, J., Newell, R.E., Parker, D.E. 1988. *Global Ocean Surface Temperature Atlas*. In final preparation for publication.

Brooks, C.F. 1926. 'Observing water-surface temperatures at sea', *Mon. Wea. Rev.* 54, 241–53.

Folland, C.K., Parker, D.E., Newman, M.R. 1985. 'Worldwide marine temperature variations on the season to century time-scale', *NOAA, Proc. 9th Annual Climate Diagnostics Workshop, Corvallis, Oregon, USA*, Washington, DC, 22–26 October 1984, 70–85.

Maury, M.F., Glaisher, J. 1852. Correspondence recorded in the Report of the *Council of the British Meteorological Society*, 25 May, 6–10.

Meteorological Office. 1881. *Management of Meteorological Instruments on Board Ship*. Printed by G.E. Eyre and W. Spottiswoode for Her Majesty's Stationery Office. Available in National Meteorological Library, Bracknell, UK.

Page, J. 1904. *Instructions to the Voluntary Meteorological Observers of the US Hydrographic Office*, Department of the Navy, Bureau of Equipment, Hydrographic Office Memo no. 119, Washington, DC, Government Printing Office. Copy also available in National Meteorological Library, Bracknell, UK.

Shearman, R.J. 1983. 'The Meteorological Office main marine data bank', *Met. Mag.* 112, 1–10.

Part II
Europe and The Mediterranean Basin

Chapter 5

Variation of air temperature and atmospheric precipitation in the region of Svalbard and of Jan Mayen

R. Brázdil   J.E. Purkyně University, Brno, Czechoslovakia

Introduction

The Norwegian Svalbard archipelago, consisting of about 1,000 islands of different sizes, with a total area of 62,400 km^2 (Greve 1985) is situated at a latitude of between approximately 74° and 81° (Figure 5.1). Typical of it is the marine Arctic climate with considerable climatic differences among the individual parts of the archipelago. Climatic conditions, together with the character of the active surface, are the main factors determining the character of the natural environment of the archipelago and processes taking place there. The main dominant features are above all different types of glacier covering about 57 per cent of the archipelago. Glaciological research activities show, however, that in the present century they have started to retreat due to changing climatic conditions.

The first continuous meteorological measurements in the region of Svalbard began at the Green Harbour station on the island of Spitsbergen as late as December 1911 (Figure 5.1). Before that time measurements had been carried out during summer or winter expeditions for periods of one year or less (their results are quoted in a paper written by Birkeland and Schon in 1938 and included in Troitsky *et al.* 1975). After 1911 continuous meteorological observations were also started at further localities listed in Table 5.1. Those observations were, of course, interrupted for different periods of time during the Second World War, and in some cases stations were shifted, which was reflected in the impairment of the homogeneity of series of some meteorological elements (such as precipitation).

In the present chapter, the objective of which is to express the main features of the variation of air temperature and atmospheric precipitation, the series of meteorological observations from the island of Jan Mayen was also considered, along with the stations of Svalbard.

Variation of air temperature in the region of Svalbard and of Jan Mayen

Steffensen (1969) completed the missing air temperature data of the war years and extended the temperature series of the Isfjord Radio station back to 1911, so that several temperature series were obtained that can be used for the analysis of long-term temperature variations. The justification for this action is confirmed by the fact that there exists a close correlation between mean monthly temperatures even at relatively distant stations, e.g. for Hornsund and Isfjord Radio the correlation coefficient is + 0.996; for Hornsund and Svalbard lufthavn + 0.994 (Baranowski 1975; Brázdil *et al.* 1987).

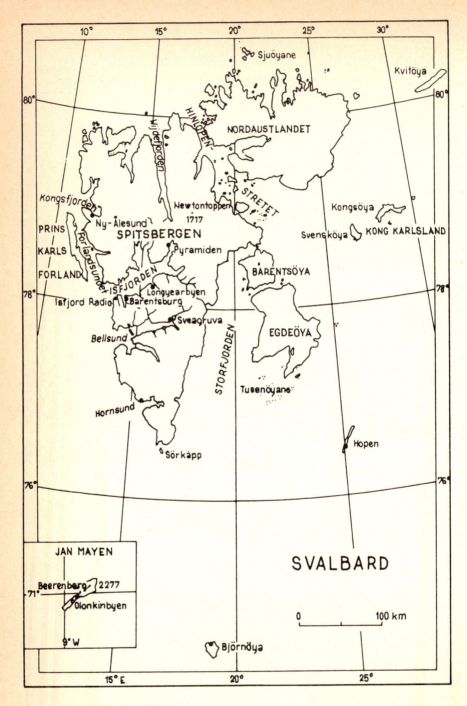

Figure 5.1 Svalbard archipelago

Note: Weather stations marked by full circles
Source: Steffensen (1982)

Table 5.1   Fundamental data about the weather stations used

Station	φ	λ	H/m	a	b
Green Harbour	78°02′N	14°14′E	11	1911	1912–30
Longyearbyen	78°13′N	15°35′E	37	1916	1957–77
Bjørnøya	74°31′N	19°01′E	15	1920	1920–85
Jan Mayen	70°56′N	8°40′W	9.5	1922	1922–85
Barentsburg	78°04′N	14°15′E	70	1930	1930–63
Isfjord Radio	78°04′N	13°38′E	5	1936	1912–76
Hopen	76°30′N	25°04′E	6	1944	1946–85
Svalbard lufthavn	78°15′N	15°28′E	28	1975	1975–85

Key:
φ   = latitude
λ   = longitude
H/m = height above sea level
a   = beginning of observation
b   = temperature series used

The variation of air temperature and/or further meteorological elements has been studied for the given region in Birkeland (1930), Ahlmann (1953), Hesselberg and Johannessen (1958), Steffensen (1969, 1982), Baranowski (1975), Kamiński (1986) etc., continued by the following analysis based on the stations in Table 5.1.

Figure 5.2 shows the course of mean annual temperatures smoothed by five-year and ten-year running averages to filter out short-term changes. From this it is evident that the stations considered are characterised by a considerable fluctuation of annual means. Thus, in the case of the Isfjord Radio station, the lowest annual mean was found in 1917, −11.0 °C, whereas the highest mean in 1954 was −2.0 °C, so that the variation range is 9.0 °C. Also at the other stations, where the very cold year of 1917 is not recorded (except for Longyearbyen, with −11.5 °C), the variation range of mean annual temperatures is relatively high: Hopen, 7.7 °C; Longyearbyen, 7.0 °C (1957–76), Bjørnøya, 5.9 °C, Jan Mayen, 4.9 °C. From Figure 5.2a the similarity of temperature variations at the Isfjord Radio, Longyearbyen and Hopen stations is evident, whereas in the case of Bjørnøya the differences are somewhat higher. Conspicuously different, as can be expected, is the variation of temperatures at Jan Mayen (thus, a continuous decrease in temperatures from 1960 to 1968 at the other stations is interrupted by two conspicuous temperature peaks).

As for the course of five-year running averages (Figure 5.2b), the minimum in the second decade of the twentieth century was followed by a significantly warmer period lasting up to practically the end of the 1950s, with three conspicuous maxima in the first half of the 1920s, in the 1930s and in the mid-1950s. The above drop in temperature culminated in the late 1960s, but the values were about 1.7 °C higher than in the minimum at the beginning of the period analysed. The subsequent rise in temperature in the first half of the 1970s did not reach the intensity of the preceding extremes, the same as the drop in the following five years. Since 1980 a new rising wave has been in progress.

In the graph of ten-year running averages (Figure 5.2c), a rising trend (warming) can be traced since the beginning of the period studied to about the mid-1930s and from the latter half of the 1960s; the falling trend (cooling) is also conspicuously expressed from the mid-1950s. Between the 1930s and the 1950s the falling trend

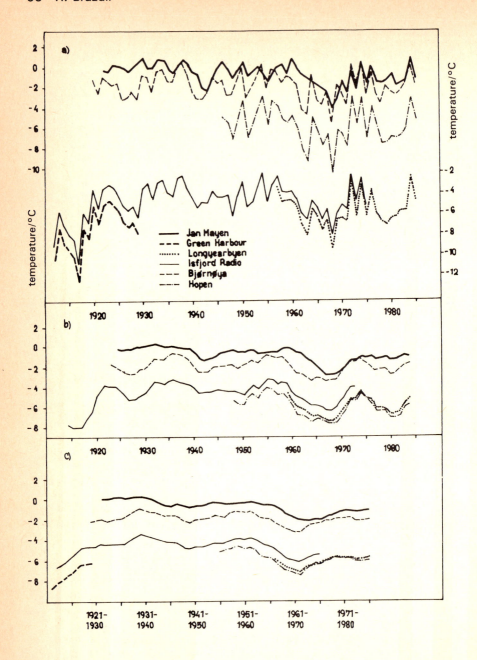

Figure 5.2   Course of mean annual air temperatures of selected stations in the region of Svalbard
(a) real values
(b) values smoothed by five-year running averages (according to Steffensen 1982, and extended to 1985)
(c) values smoothed by ten-year running averages.

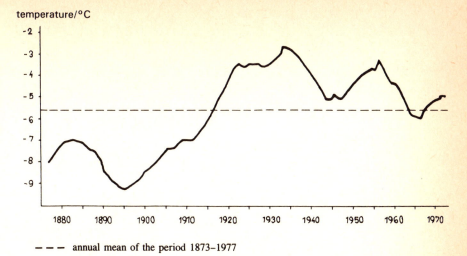

temperature/°C

─ ─ ─   annual mean of the period 1873–1977

Figure 5.3   Course of mean annual air temperatures at Spitsbergen smoothed by ten-year running averages

Source: Data by Markin and the Isfjord Radio Station (Kamiński 1986)

changed into a rising one.

For comparison, the course of a series is given in Figure 5.3, smoothed by ten-year running averages. That series includes data calculated by Markin from discontinuous measurements in 1873–1912 and the data of the Isfjord Radio station for the years 1913–77 (Kamiński 1986). The above information coincides to some extent with the description of global changes in air temperature of high geographical latitudes, such as those given in Kelly *et al.* (1982) and Vinnikov (1986).

As for the course of seasonal air temperatures, this generally exhibits features identical with the described course of annual values, the winter means exhibiting the maximum amplitude and the summer months the minimum (thus, at Isfjord Radio the variation range of winter temperatures is double that of the summer). That is also clearly seen in the course of temperatures of the individual months in the example of the stations Isfjord Radio and Svalbard lufthavn (Figure 5.4).

Variation of atmospheric precipitation in the region of Svalbard and of Jan Mayen

The description of the variation of atmospheric precipitation in the region of Svalbard poses several problems. Precipitation is a meteorological element which is characterised by great time and spatial variabilities, which makes it practically impossible to complete the missing data from the war years with the required accuracy. The spatial distribution of the precipitation is, moreover, made rather complicated by the complexity of the active surface in the region of Spitsbergen proper, above all by orographic conditions. The possible changes of the station's position will be reflected rather in the inhomogeneity of the precipitation series than

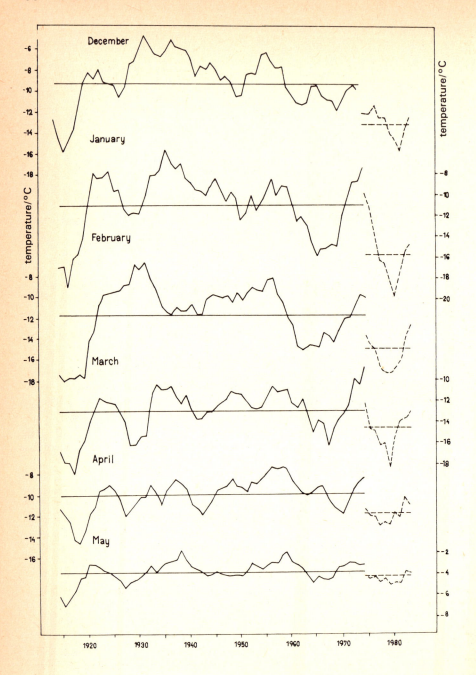

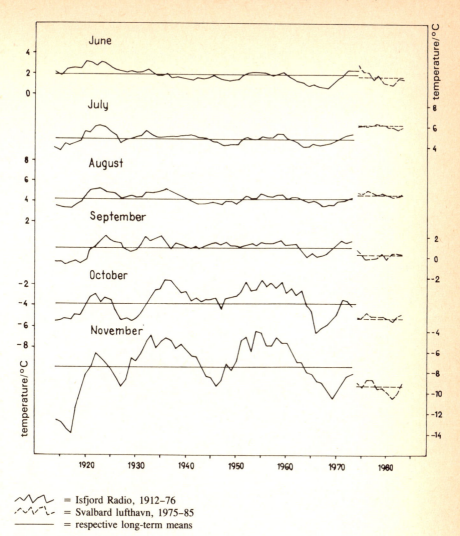

Figure 5.4 Course of mean monthly air temperatures smoothed by five-year running averages for the Isfjord radio station and the Svalbard lufthavn station

in the temperature record. It is also necessary to remember that there can be considerable error in measuring precipitation, which is frequently accompanied by a strong wind or drifting snow. For those reasons long-term means of stations not very far from each other can differ significantly. Thus, according to Pereyma (1983) and Steffensen (1982) the annual precipitation sums of stations in the central part of Isfjorden are usually about between one-half and one-third lower than those of the western part of the island.

With respect to the above statement, with the exception of the Hopen and Bjørnøya stations (1946–85) we have practically no longer precipitation series that might be considered at least relatively homogeneous. The variation of precipitation

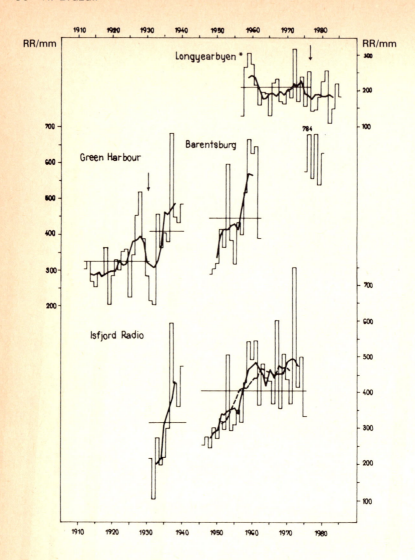

* from 1977 onwards the data of the station Svalbard lufthavn have been used

―――――― = five-year running averages

- - - - - = ten-year running averages

――――▶ = mean points where the homogeneity of the given series is interrupted

Figure 5.5 Course of annual precipitation sums smoothed by five-year and ten-year running averages of selected stations in the region of Svalbard.

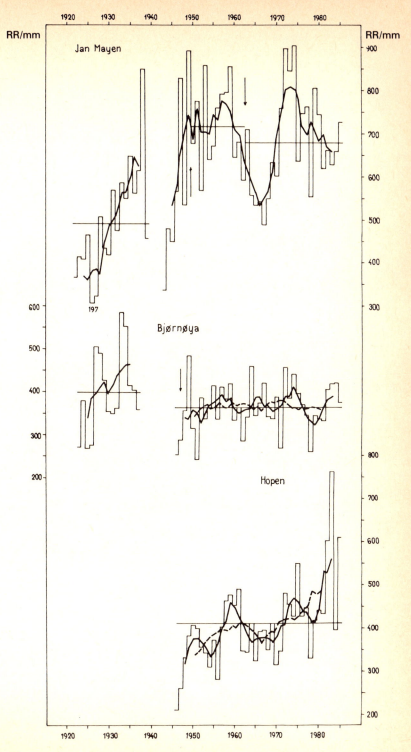

can thus be reconstructed only on the basis of changes in shorter time intervals. According to the course of five-year running averages at the Green Harbour, Isfjord Radio, Barentsburg, Longyearbyen, and Svalbard lufthavn stations (Figure 5.5) it can be judged that in the central part of the island of Spitsbergen (the Isfjorden area), after low sums in the second decade of the twentieth century there came a maximum in the latter half of the 1920s and afterwards a drop in precipitation at the beginning of the following decade in the latter half of the 1930s. It is difficult to judge the relative amount of those extremes with respect to each other, even though it can be judged that the latter maximum was more conspicuous. The latter half of the 1940s was characterised by below-average annual sums lasting until 1958 under a conspicuous rising trend of precipitation. A precipitation maximum was reached about the year 1960. Another rise, after the drop in precipitation in the 1960s, came after 1970. In recent years (according to data of the Svalbard lufthavn station) the annual sums have run a more or less balanced course.

As for the remaining part of Svalbard and of Jan Mayen (Figure 5.5), the 1920s and the first half of the 1930s were characterised by a more or less continuous increase in precipitation. After the Second World War the maxima at all three stations fell until the latter half of the 1950s and in the first half of the 1970s; below-average annual precipitation of the 1960s was not reflected at Bjørnøya. Whereas at Jan Mayen the precipitation, after the maximum in the first half of the 1970s, had a falling trend, at Bjørnøya and Hopen this trend was interrupted before 1980 and replaced by precipitation increase, particularly conspicuous at Hopen.

Cyclicity of air temperature and atmospheric precipitation in the region of Svalbard and of Jan Mayen

Hitherto numerous studies of time changes in temperature and precipitation series have shown that it is practically impossible to find any periodicity in them in the mathematical sense of the word, i.e. for $x(t_i) = (t_i + T)$ to hold, where $x(t_i)$ is the time series ($i = 1, 2, \ldots$) and T the period. Of temperature and particularly precipitation series, characterised by a high component of noise suppressing the climatic signal proper, it is typical that periods included in them change their phases and amplitudes, one periodical oscillation being able to pass into the next, or change into an aperiodic one after some time. The variation of temperatures and precipitation has a quasi-periodic character for which the classification of Schönwiese (1974) uses the term 'rhythm' and for which Drozdov and Grigoryeva (1971) use the term 'cyclicity'.

For the analysis proper, temperature and precipitation series were chosen which had minimum lengths of 40 years and which could thus be considered homogeneous, i.e. temperature series for the Isfjord Radio, Hopen, Bjørnøya, Jan Mayen stations, and precipitation series for the Hopen and Bjørnøya stations. These series were analysed by spectral methods analysis after Blackmann and Tukey (B&T), maximum entropy spectral analysis (MESA) and dynamical MESA, described in the form used in the present chapter by Brázdil (1987).

On the basis of spectral analysis after B&T, significant periods for air temperature seem to be those of 2.0 years for Isfjord Radio, 2.5 years for Bjørnøya, and 2.0 years for Jan Mayen. The importance of the period in the length of 2.5 years is also confirmed by its insignificant occurrence for Hopen as well as its

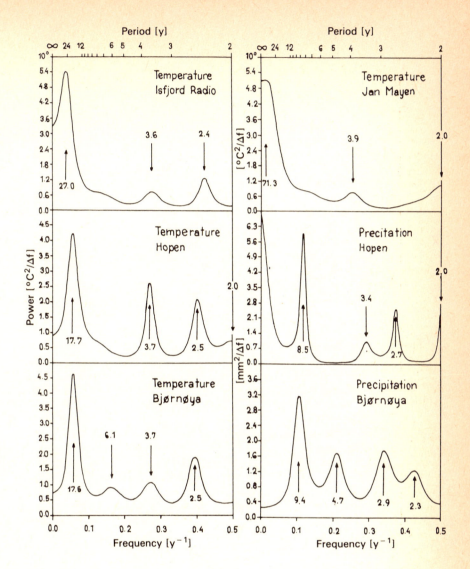

Figure 5.6 Variance spectra of series of mean annual air temperatures and annual precipitation sums of selected stations in the region of Svalbard

Note: MESA, $M = 10$, number of frequencies 500. For periods of observations, see Table 5.1.

determination by the MESA method which, in comparison with the B&T method, is characterised by a better distinctive ability (Figure 5.6). Isfjord Radio, Hopen and Bjørnøya also exhibit a period of 3.6–3.7 years (when analysed by the B&T method it corresponds to an insignificant period of 3.6 or 4.0 years), Jan Mayen 3.9 years. An important component of variance spectra are also long-period oscillations. For stations Hopen and Bjørnøya they have a length of 17.6–17.7 years, for

Period/y

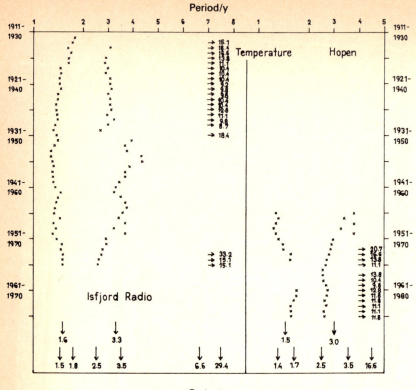

Period/y

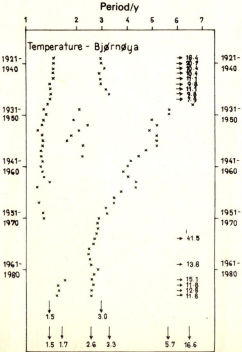

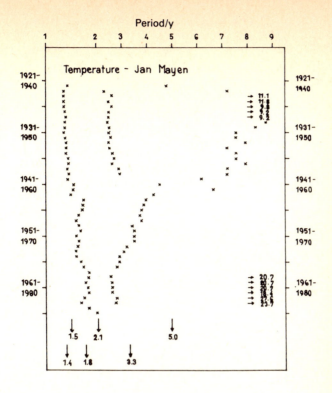

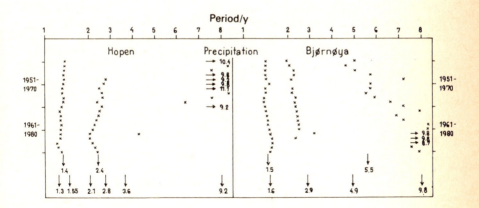

Figure 5.7 Dynamical MESA of series of mean monthly air temperatures and monthly precipitation sums of selected stations in the region of Svalbard

Note: Numbers in the first line mark medium periods calculated by MESA for $M = 50$ and $M = 200$ frequencies, in the second line for $M = 100$ and $M = 500$ frequencies.

Isfjord Radio 27.0 years, and for Jan Mayen as much as 71.3 years.

As for the spectra of precipitation series, they are distinctly different both from those of temperature and from one another. By the B&T method none of the periods found is significant and the two series have only one common period — that of 2.9 years (for Hopen the corresponding length of the period according to MESA is 2.7 years). In the variance spectra themselves, besides the trend for Hopen there are also marked periods in the length of 8.5 years for Hopen and 9.4 years for Bjørnøya.

The periods found need not, of course, be of the same length and intensity in the course of the whole period studied. Therefore dynamical MESA was applied to temperature and precipitation series. The maxima of corresponding spectra are shown in Figure 5.7 which indicates a relatively considerable inconstancy of the individual periods in length, in the sense of both shortening and lengthening of the given oscillation, the given oscillation not asserting itself in certain time periods. Relative stability of length appears for the fluctuation at the interval of 1–2 years. In the case of temperature series, with the exception of Hopen, the shortening of the length of oscillations is well seen from the beginning of the period studied to its end, e.g. for Bjørnøya from 10–20 years to 2.5–2.6 years, for Jan Mayen from 9–12 to 2.6 years. At the Isfjord Radio station, in the first 20 years apart from the periodicity of about 3 years there also appeared an oscillation within 9–18 years; then the long-term component practically disappeared and the medium-period component was shortened from about 4.4 years to only 2.6 years. From Figure 5.7 it can also be seen that periods stated by MESA for the whole series are a certain 'averaging' of different periods appearing in partial time spaces (such as the periods of 2.1 and 5.0 years for Jan Mayen). In the case of precipitation series, of interest is the lengthening of the medium period of 5.5 years for Bjørnøya from 4.6–5.0 years to 7.5–10.0 years.

As in most other papers of this type dealing with the analysis of variation of air temperature and precipitation, the clarification of causes of those fluctuations remains problematic. Whereas the most distinct annual period appearing in the series of monthly values is a reflection of the seasonal cycle given by the rotation of the earth around the Sun, the further oscillations found can be due to the cyclicity of processes of atmospheric circulation and also the internal instability of the climatic system.

Conclusion

The described variations of temperatures and precipitation in the region of Svalbard are in good correlation with the knowledge of the changes in Spitsbergen glaciers. According to Baranowski (1975) the annual balance of the mass of glaciers depends mainly on summer temperatures and winter precipitation. Koryakin (writing in 1974, quoted in Jania 1982) states for the region of Hornsund that after a slow retreat of glacier snouts in the period 1900–18 there followed a marked increase in the rate of recession and, after a slowing down of the rate in 1938–49, the recession increased again in the 1950s. According to Baranowski (1977a; 1977b) from 1870 onwards, the Spitsbergen transition glaciers changed in area at the following rates: from 1870 to 1900 by +2.7 per cent, from 1900 to 1936 by −2.1 per cent, and from 1936 to 1950 by −1.0 per cent. Global data about the retreat of glaciers in the region of Svalbard witnessing the warming trend of climate are given by

Koryakin (in Kotlyakov 1985). Thus, in the southern part of the island of Spitsbergen the area of glaciers decreased by 11.7 per cent (1900–76), in the north-western by 6.8 per cent (1907–76), in the central by 6.0 per cent (1912–66), and in the north-eastern by 4.2 per cent (1900–76).

From a more global view temperature oscillations are on the whole in good agreement with data concerning the area of sea ice in the Arctic. Thus, according to Vinnikov *et al.* (1980) who started from the area of ice in the Arctic Ocean (including its peripheral seas) from 1946, after the reduction of its extent in the mid-1950s the area increased and in 1967 a maximum was reached in the 30 years studied. In the late 1960s the area of sea ice was decreasing, this process continued also in the 1970s.

The analysis has shown that, while for the descriptions of air temperature changes existing measurements represent, with respect to a larger areal range of temperature anomalies, a relatively satisfactory starting material, from the point of view of studying variation of atmospheric precipitation this starting assumption is not fulfilled. Despite that, it is possible to point to the chief characteristics of variations of air temperature and precipitation in the region of Svalbard and of Jan Mayen and to the regional peculiarities of these fluctuations.

The above analysis will need to be supported also by a dynamic climatological and/or synoptic-climatological analysis of the fluctuations described, the reason for which is to be sought in the variability and variation of circulation conditions in this region.

Acknowledgements

The author would like to thank Bjørn Aune from Det Norske Meteorologiske Institutt in Oslo for kindly granting temperature and precipitation data from the region of Svalbard and of Jan Mayen.

References

Ahlmann, H.W. 1953. 'Glacier variations and climatic fluctuation', Bowman Memorial Lectures, *Amer. Geogr. Soc.*, ser. 3.

Baranowski, S. 1975. 'The climate of West Spitsbergen in the light of material obtained from Isfjord Radio and Hornsund', *Acta Universitatis Wratislaviensis, Spitsbergen Expeditions* I, 251, 21–34.

Baranowski, S. 1977a. 'The subpolar glaciers of Spitsbergen seen against the climate of this region', *Acta Universitatis Wratislaviensis* 410, Wrocław.

Baranowski, S. 1977b. 'Changes of Spitsbergen glaciation at the end of the Pleistocene and in the Holocene', *Quaestiones Geographicae* 4, 5–27.

Birkeland, B.J.1930. 'Temperaturvariationen auf Spitzbergen', *Meteorologische Zeitschrift* 47, 6, 234–6.

Brázdil, R. 1987. 'Variation of atmospheric precipitation in the C.S.S.R. with respect to precipitation changes in the European region', *Folia Fac. Sci. Nat. Univ. Purk. Brun., Geographia*, Brno.

Brázdil, R., Chmal, H., Kida, J., Klementowski, J., Konečný, M., Pereyma, J., Piasecki, J., Prošek, P., Sobik, M. Szczepankiewicz-Szmyrka, A. 1987. 'Results of Investigations of the Research Geographical Expedition Spitsbergen 1985', Folia Fac. Sci. Nat. Univ. Purk. Brun., Geographia, Brno.

Drozdov, O.A., Grigoryeva, A.S. 1971. 'Mnogoletniye tsiklitcheskiye kolebaniya atmosfernykh osadkov na territorii SSSR', *Gidromet. izd.*, Leningrad.

Greve, T. 1975. *Svalbard. Norway in the Arctic Ocean*, Grøndahl & Søns Forlag A/S.

Hesselberg, T., Johannessen, W.T. 1958. 'The recent variations of the climate at the Norwegian Arctic stations' in *Polar Atmosphere Symposium Part 1, Meteorological Section*, London, Pergamon Press.

Jania, J. 1982. 'Ablacja przez ''cielenie'' i wycofywanie sie lodowcáw Hornsundu (Spitsbergen) w XX wieku. Wyniki wstepne', *Wyprawy Polarne Uniwersytetu Ślaskiego 1977–1980*, 1, 13–46.

Kamiński, A. 1986. 'Wspólczesne wahania temperatury powietrza na Spitsbergenie Zachodnim', *II Zjazd Geografów Polskich, Lódź*, 58–62.

Kelly, P.M., Jones, P.D., Sear, C.B., Cherry, B.S.G., Tavakol, R.K. 1982. 'Variations in surface air temperatures: Part 2. Arctic region, 1881–1980', *Mon. Wea. Rev.* 110, 71–83.

Kotlyakov, V.M., ed. 1985. *Glatsiologiya Shpitsbergena*, Moscow, Nauka.

Pereyma, J. 1983. 'Climatological problems of the Hornsund area, Spitsbergen', *Acta Universitatis Wratislaviensis* 714, Wrocław.

Schönwiese, C.D. 1974. 'Schwankungsklimatologie im Frequenz- und Zeitbereich', *Wiss. Mitteil. Univ. München*, Meteorol. Inst., 24, Munich.

Steffensen, E.L. 1969. 'The climate and its recent variations at the Norwegian Arctic stations', *Meteorologiske Annaler* 5, 8.

Steffensen, E.L. 1982. 'The climate at Norwegian Arctic stations, Oslo, Det Norske Meteorologiske Institutt.

Troitsky, L.S., Zinger, E.M., Koryakin, V.S., Markin, V.A. Mikhailov, V.I. 1975. *Oledeneniye Shpitsbergena (Svalbarda)*, Moscow, Nauka.

Vinnikov, K.Y. 1986. *Tchustvitelnost klimata*, Leningrad, Gidrometeoizdat.

Vinnikov, K.Y., Gruza, G.V., Zakharov, V.F., Kirillov, A.A., Kovyneva, N.P., Rankova, E.Y. 1980. 'Sovremennyye izmeneniya klimata Severnogo polushariya', *Meteorologiya i gidrologiya* 6, 5–17.

Chapter 6

Variations in the wind resource over the British Isles for potential power production

T.D. Davies, J.P. Palutikof, T. Holt and P.M. Kelly
University of East Anglia, United Kingdom

Introduction

A recent report by the British Wind Energy Association (1987) claims that, 'on good, windy sites' wind turbines can generate electricity at lower cost than coal, oil, nuclear or other renewable plant. The BWEA asserts that 20 per cent of all generation is a reasonable initial target. Around 75 per cent of the world's commercial wind energy capacity is located in wind farms in California (~ 1000 MW), although a number of single turbines have been constructed or are being commissioned in the UK. For a review of the current distribution of wind turbine generating capacity, see Palutikof *et al.* (1987a).

Since wind turbines have a design life of around 30–40 years, it is necessary to consider the variability of the wind resource on the time-scale of years to decades. Although amenity and transmission considerations may, in practice, be overwhelming in the choice of site, it is clearly important to identify candidate sites where the wind frequency distribution has a high mean value and a low variability. In addition, an appreciation of the possible range of windspeed frequency distributions at a site (at wind turbine hub height) is necessary so that a turbine with an appropriate power curve can be installed. If the cut-in and rated speeds are not appropriate, much of the potential power will not be realised.

Wind energy engineers are interested in the resolution of the wind regime to all time-scales, with the greatest interest thus far being shown in the hourly-daily scale, with the tacit assumption that the frequency distributions do not exhibit trend on time-scales of years to decades. In this chapter, we shall examine this aspect of the wind resource which is almost always overlooked. We shall address the issue in terms of annual or monthly mean wind speeds in order to consider the links with climatological-scale atmospheric circulations. We shall not consider, here, any changes in frequency distributions on the daily- or hourly-scale which may occur over years to decades, although these are of ultimate importance for the exploitation of the wind resource in the UK.

Spatial variability

Figure 6.1 shows the distribution of the mean wind speed over the UK for the 1956–82 period. The data network was chosen after stringent quality-control and homogeneity-checking procedures; many other records were rejected during this vetting (Palutikof *et al.* 1984; 1985a). The data were corrected to a common effective height of 10 m, using a power law exponent of 0.16 (Halliday and Lipman 1982). Other stations in the Scottish islands (two), Ireland (one) and the Channel

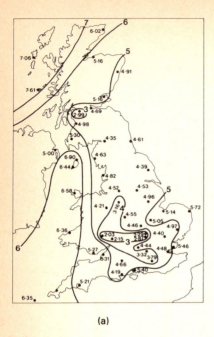 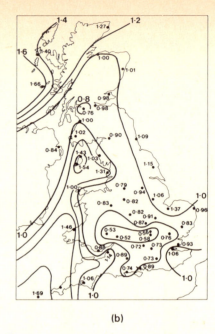

(a)   (b)

Figure 6.1   (a) Annual mean wind-speed (m s^{-1}) and (b) annual standard deviations (m s^{-1}) over the period 1956–82

Note: Four stations outside the map area were also used to aid the construction of the isopleths

Islands (one) were also utilised for the construction of the isolines. Spatial distributions such as these indicate possible favourable regions for the location of wind turbines, although there are severe problems in extrapolating the mean wind regime from observing sites to candidate sites (Palutikof *et al.* 1987a). However, identification of regions with relatively high wind speeds and relatively low variability is a useful first step. Much of the east coast experiences relatively low wind speeds (less than 5 m s^{-1}) and a standard deviation greater than 1 m s^{-1}, indicating, on the face of it, an unfavourable degree of variability for wind power production. However, siting strategies might take account of the synoptic scale of circulation systems to ensure that a turbine network is designed to minimise common periods of non-production because of low wind speeds. Such strategies might involve spatially diverse siting (e.g., Anderson *et al.* 1987).

Temporal variability

Spatial variability, over a common period (Figure 6.1), is relatively easy to consider. More problematical is temporal variation in the wind regime, particularly when there may be a considerable spatial variation in the response over time. Figure 6.2 shows seven long-term wind-speed records from stations in the British Isles. These records were relatively easy to compile, using mainly published sources. Details of each of the seven sites and the records are available in Palutikof

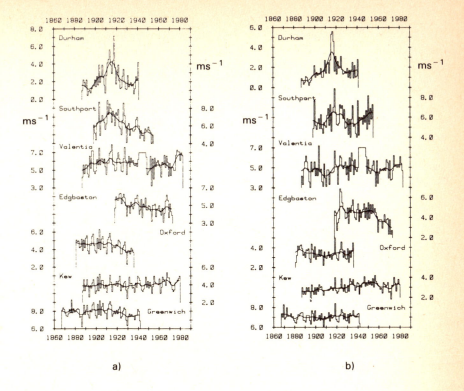

a)   b)

Figure 6.2   Mean wind speeds (m s^{-1}) for seven long-term stations: (a) March;
(b) September

et al. (1985a). The months of March and September are representative of
conditions in the winter and summer half-years (Palutikof *et al.* 1985b) and both
months exhibit similar curves at each station (Figure 6.2). Two distinct regimes are
represented in the analysis of the March observations; the two northern stations
(Durham and Southport) show a marked peak in the early part of this century and
clear trends are exhibited over periods of decades. The other five records (central
and southern England, and the west coat of Ireland) present a different picture.
Long-term trends are small or non-existent. For the September analysis, Durham
and Southport again display peak values in the 1910s, although the highest wind-
speed pentade is 1915–19. The lowest wind-speed pentades are the same as for
March at both stations. The recovery in the downward trend which took place in
March at about 1940, but was not sustained, persists to the end of the Southport
record in September. The remaining stations do not show any marked and coherent
trend in September, although the downward movement at Edgbaston is somewhat
more marked than in March.

Power production

In site selection for wind turbines, it is desirable to consider the longest-available wind speed records, in order to have the best possible estimate of the range of variability of the future wind regime. It also needs to be borne in mind that potential power production from the wind is not linearly proportional to wind speed. The available power is proportional to the cube of the wind speed (Palutikof *et al.* 1985b). The immediate implications of this are shown in Figure 6.3 which shows the annual mean wind-speed record for Southport, together with the annual mean values cubed. So, the potential power output varied by a factor of ~ 3.0 from the year of the lowest annual mean wind speed to the year of the highest annual wind speed. The appropriate factor is ~ 1.6 for the smoothed curve in Figure 6.3. If estimates of the power available from any proposed wind turbine had been based

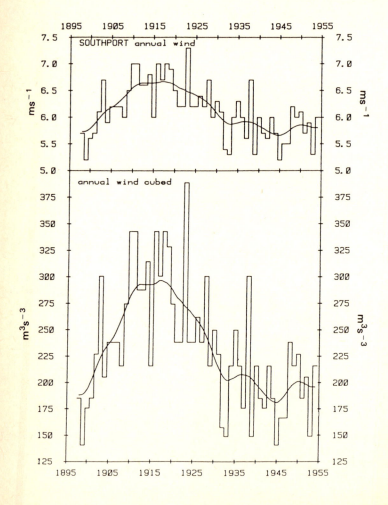

Figure 6.3 Annual mean wind speeds (m s^{-1}), and mean wind speed cubed, for Southport

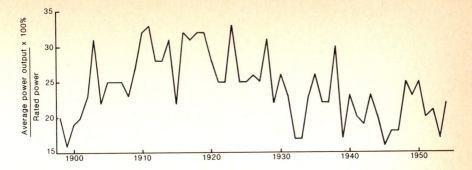

Figure 6.4   Time series of average power output as a percentage of rated power from a hypothetical wind turbine (see text) sited at Southport, 1898–1954

on data for the 1910–19 decade, the output would have been disappointing in the 1940s.

It can be misleading to discuss power production in terms of a function of wind-speed cubed, since actual output depends on turbine characteristics as well as the wind regime (Palutikof *et al.* 1987a). A particular turbine will start to produce energy at a *cut-in speed*, the output will increase up to the *rated wind speed*, beyond which it will remain constant to the *cut-out wind speed*, beyond which the turbine will be closed down to avoid structural damage. The actual energy production from a turbine is therefore a function of these individual turbine characteristics and the wind-speed frequency distribution (and so, moreover, the use of long-period *mean* wind speeds will not provide exact assessments of the *actual* power production (Edwards and Dawber 1980) but, for the purposes of the discussion, we shall continue to use annual mean wind speeds). Cliff (1977) presented a set of curves relating wind turbine performance characteristics and observed mean wind-speed characteristics to power output. We have utilised Cliff's calculations using the characteristics of a turbine currently in operation in the Orkneys (Howden 300 KW, rated wind speed 16.7 m s^{-1}, cut-out speed 28 m s^{-1} and a hub-height of 22 m), and the Southport wind record (Figure 6.3) to estimate the average power output (as a percentage of the rated power) of a similar turbine located at Southport. Hub-height wind speeds were adjusted to the effective height of the Southport anemometer (13 m) using the power law equation with an exponent of 0.16 (Halliday and Lipman, 1982) giving a 13-metre-rated speed of 12 m s^{-1} and a 13-metre cut-out speed of 20 m s^{-1}. Figure 6.4 shows the calculated time series of annual average power output from the wind turbine hypothetically sited at Southport. For the maximum annual mean the average power output would be 33 per cent of the rated output, whereas for the minimum year it would be 16 per cent.

It is instructive to consider how alterations in the turbine characteristics can be used to minimise the impact of temporal variability. For example, a rated speed of 7 m s^{-1} and a cut-out speed of 12 m s^{-1} produce variations in average power output, for the extreme individual years, from 56 to 46 per cent. Judicious matching of the turbine characteristics with the observed wind regime (over the longest possible period) can maximise the return from the original economic cost.

Links with the regional circulation

Analysis of a denser network (than that discussed in respect of Figure 6.2) of anemometer sites from the mid-1950s showed that many locations experienced a marked decline in wind speeds from the 1960s to the 1970s (Palutikof *et al.* 1985b). Jones and Kelly (1982) applied principal components analysis to examine the major changes in atmospheric circulation over the British Isles, using Lamb Weather types (Lamb 1972), which are regarded as succinct descriptions of the daily synoptic pattern. Jones and Kelly identified the 1970s as a decade of increasing frequency of anticyclonic conditions (at the expense of westerly conditions) when compared to the 1960s. Their finding is entirely consistent with the observed wind-speed records. The Lamb Weather Type principal components analysis produced a first component which accounted for 34 per cent of the total variance in the data set (1861–1980), and which was most strongly loaded by anticyclonic (A, positive) and westerly (W, negative) weather types.

Since the pattern of observed wind speeds over Britain shows considerable spatial variability (Figure 6.1), it becomes quite a complex problem to analyse temporal variability for the whole region (Palutikof *et al.* 1987a). This problem has been tackled using principal components analysis on a data matrix of 42 stations (spread as uniformly as possible over the UK) with 20 years (1962–81) of mean monthly wind-speed values. The data were expressed as anomalies from the long-term monthly mean, thus effectively excluding the seasonal component from the analysis. Full details are available in Palutikof *et al.* (1987a). The first principal component accounted for 51 per cent of the variance in the data set. The loadings of this component were of the same sign and approximately the same magnitude at the great majority of stations, indicating that the first component summarised the broad-scale characteristics of the wind field over Britain. From the monthly scores of the first principal component, a time series of the annual means of the 12 monthly values was generated. A correlation analysis between this series and the annual first component scores for the principal components analysis of the Lamb Weather Types (Jones and Kelly 1982) over the period 1962–80 produced a coefficient of -0.9 (significant at the 1 % level). This confirms that generally high wind speeds over Britain are associated with a predominance of westerly days at the expense of anticyclonic days, and vice versa. Thus, long-term variability in the regional-scale atmospheric circulation patterns influences the local wind field at the same time-scales (Palutikof *et al.* 1987a).

However, all areas of Britain may not exhibit similar atmospheric circulation–wind–speed relationships. In order to examine possible spatial differences in the response, a monthly A/W index was derived from the number of anticyclonic and westerly days in any given month weighted by the loadings of the two weather types on the first principal component (Palutikof *et al.* 1987a). A time series for the period 1956–82 was constructed, which expresses the relative importance of A and W events. For the stations shown in Figure 6.5, the wind-speed time series for each month was correlated with the A/W index, giving 12 correlation coefficients. The number of months at each station with significant correlations at the 5 % level (allowing for possible autocorrelation — Brooks and Carruthers 1953; Mitchell *et al.* 1966) is shown in Figure 6.5. There is a clear N–S alignment in the distribution, with higher values in the west. West coast wind speeds exhibit close relationships with the relative predominance of westerly conditions. On the east coast, there are other mechanisms of equal or greater importance, which may include sea-breeze circulations in these relatively sheltered locations.

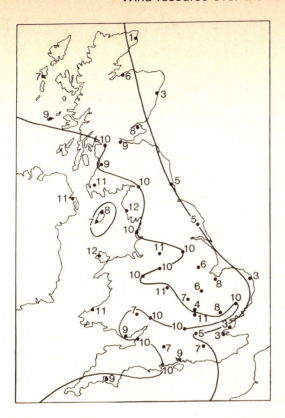

Figure 6.5 Number of months (maximum of twelve) in which wind speed correlates significantly with A/W index (see text)

Links with the hemispheric-scale circulation

Since the Lamb Weather Type classification is restricted to a geographical area approximating to the British Isles, it does not provide direct information about the links between wind regimes and the atmospheric circulation on a larger space-scale. To this end, a set of 'high' wind-speed and 'low' wind-speed months was identified using the results of the principal components analysis of the wind-speed records. Arbitrary levels of positive and negative scores for the first principal component were set (Palutikof *et al.* 1987a). This identified six 'high' wind-speed months and eight 'low' wind-speed months over the 1961–80 period. Then, for each month in the two samples, using data supplied by the UK Meteorological Office, the sea-level pressure anomaly (from the long-term 1899–1980 mean for that month) was calculated, on a 5° × 10° longitude grid, for the area shown in Figure 6.6. For the two sets of 'high' wind-speed and the 'low' wind-speed months, the anomalies were then summed algebraically to produce the two composite pressure-anomaly maps. Figure 6.6a shows the enhancement of the westerly wind component during the 'high' wind-speed months, whereas the 'low' wind-speed months are related to a relative decrease in the zonal gradient. This analysis confirms that changes in the wind field over the British Isles are produced by variations in the atmospheric

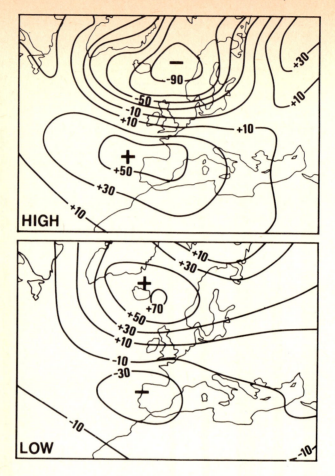

Figure 6.6 Sea level pressure anomalies (tenths of mb) from the 1899–1980 mean for 'high' and 'low' wind speed months (see text)

circulation at the hemispheric scale. This is a useful connection, since more is known about large scale variations in the pressure field over periods of up to ~ 100 years, than is known about changes in the wind field on local or national scale.

Conclusions

We have demonstrated that temporal variability in mean annual wind speeds is an important factor in any study of wind-power potential. Variations over time have implications not only for overall resource assessment and economic planning, but also for matching the characteristics of any proposed turbine to the site wind regime.

Trends in monthly mean wind speed are related to indicators of the daily variations in atmospheric circulation on a sub-regional scale, as reflected in the Lamb

Weather Type classification. This confirms that the use of monthly wind-speed statistics is of direct relevance to the wind energy industry (which, operationally, is more concerned with wind speeds at the daily, and higher-resolution, level), since monthly mean wind speeds represent the aggregation of daily conditions. Of particular interest to the industry, however, are the precise ways in which daily, hourly or shorter-period frequency distributions respond to the sub-regional circulation changes which produce the observed variations in the monthly mean wind speed (Palutikof *et al.* 1987b).

The variations in mean monthly or annual wind speeds which are related to the sub-regional variations in atmospheric circulations also exhibit a demonstrable connection with changes in the sea-level pressure distributions on a hemispheric scale.

Acknowledgements

Financial support is gratefully acknowledged from UK Department of Energy, the Science and Engineering Research Council and the Central Electricity Generating Board.

Reference

Anderson, M.B., Newton, K., Ryle, M., Scott, P.F. 1987. 'Short-term storage and wind power availability', *Nature* 275, 432–4.

British Wind Energy Association 1987. *Wind Power for the UK*, London BWEA.

Brooks, C.E.P., Carruthers, N. 1953. *Handbook of Statistical Methods in Meteorology*, London, HMSO.

Cliff, W.C. 1977. *The Effect of Generalised Wind Characteristics on Annual Power Estimated from Wind Turbine Generators*, PNL-2436, Battelle Pacific Northwest Laboratories, Richland, WA.

Edwards, P.J., Dawber, K.R. 1980. 'An investigation of wind energy prospects in the Otago region of New Zealand', *J. Ind. Aerodynamics* 5, 281–96.

Halliday, J.A., Lipman, N.H. 1982. 'Wind speed statistics of 14 widely dispersed UK meteorological stations', *Proc. 4th BWEA Conference*, BHRA Fluid Engineering, London, 180–8.

Jones, P.D., Kelly, P.M. 1982. 'Principal components analysis of the Lamb catalogue of daily weather types: Pt. 1, annual frequencies', *J. Climatol* 2, 147–57.

Lamb, H.H. 1972. 'British Isles weather types and a register of the daily sequence of circulation patterns, 1861–1971', *Geophys. Mem.* 116.

Mitchell J.M., Dzerdzeevskii, B., Flohn, H. Hofmeyr, W.L., Lamb, H.H., Rao, K.N., Wallén. C.C. 1966. 'Climatic Change', *WMO Tech. Note 79*, WMO no. 195, World Meteorological Organisation, Geneva.

Palutikof, J.P., Davies, T.D., Kelly, P.M. 1984. 'A databank of wind-speed records for the British Isles and offshore waters' in P. Musgrove (ed.), *Wind Energy Conversion 1984*, Cambridge, Cambridge University Press, 414–25.

Palutikof, J.P., Davies, T.D., Kelly, P.M. 1985a. 'An analysis of seven long-term wind-speed records for the British Isles with particular reference to the implications for wind power production' in A. Garrad (ed.), *Wind Energy Conversion 1985*, London, Mechanical Engineering Publications, 235–40.

Palutikof, J.P., Kelly, P.M. Davies, T.D., Halliday, J.A. 1987a. 'Impacts of spatial and temporal windspeed variability on wind energy output', *J. Climatol. Appl. Meteorol.* 269,

1124–33.

Palutikof, J.P., Davies, T.D., Kelly, P.M., Halliday, J.A., 1987b. 'Temporal and spatial variations in hourly wind speeds over the British Isles' in W. Palz and E. Sesto (eds) *Proc. European Wind Energy Association, Rome 1986*, Rome, A Raguzzi, Bookshops for Scientific Publications, 231–6.

Chapter 7

Precipitation composition and climatic change in Scotland and southern Scandinavia

T.D. Davies, G. Farmer, G.M. Glover*, R. Barthelmie, P.M. Kelly and P. Brimblecombe University of East Anglia and *CEGB, Leatherhead, United Kingdom

Introduction

There are strong links between the chemical composition of individual precipitation episodes and meteorology (see, e.g., Smith and Hunt 1978: Davies *et al.* 1988). Trajectory path is one important factor (urban/industrial, marine sources, etc.), but the associated conditions are also relevant. As an example, some snowfalls which occur in the Scottish highlands in association with easterly trajectories are highly polluted; others which are associated with similar trajectories are relatively clean. The critical feature appears not to be the plan of the back-trajectory, but the presence of elevated subsidence inversions which allow long-range transport of pollution with greatly inhibited dispersion. The links between climate and wet pollution deposition over periods of months or years are less clearly defined and few studies have attempted to address them (one example is Davies *et al.* 1986). There have been observed trends in wet acidic deposition over Europe over years to decades, with regional variations (Rodhe and Granat 1984). Variations in emissions may have produced these observed trends, but long-term changes in climate could also have played a role. It is reasonable to postulate that, since a precipitation event is frequently the end-product of atmospheric processes which have taken days to evolve, the character of the climate over a period of (say) a month at a particular location will be reflected in the wet pollution deposition over a similar period at that same location.

In this chapter we shall examine the links between wet acidic deposition at one location in Scotland and a weather classification system at the daily level. The rationale for this study is that, if such a link can be established, weather classification systems may be used as a surrogate for trajectory analysis in studies over longer periods (years to decades). Moreover, when the wet deposition is collected on a monthly basis, it is not possible to determine the appropriate trajectories.

Wet acidic deposition and daily weather types

Eskdalemuir, in southern Scotland (55° 19′N, 3° 12′W), is one of the stations in the EMEP precipitation chemistry network (Schaug *et al.* various years), where chemical composition of precipitation collected on a daily basis is determined. On each day when rain fell, the prevailing weather type (Lamb 1972a) was identified. Lamb Weather Types (LWTs) provide a succinct characterisation of the synoptic circulation and the record extends back to 1861, allowing studies of long-term trends. On an annual basis, three LWTs account for much of the variation in the circulation over the UK (Jones and Kelly 1982): westerly (W): anticyclonic (A): and cyclonic (C). A-types produce little precipitation, although anticyclones located over Europe have been demonstrated to be of eventual importance for acidic deposition in distant receptor areas (Smith and Hunt 1978). C- and W-types produce more than 50 per cent of the precipitation in the UK, and no other type (of which there are 27 in total) accounts for more than about 10 per cent.

Trajectory analyses have been undertaken for precipitation events occurring at Eskdalemuir during C- or W- days. The synoptic situation associated with the C-type classification is such that, when there is frontal precipitation over Scotland, back-trajectories often, but not exclusively, originate over England or near-continental Europe. This may be expected to lead to relatively acidic precipitation over Scotland (Fowler and Cape 1984). W-type precipitation, on the other hand, is more commonly associated with maritime trajectories and less-polluted rainfall in Scotland (Fowler and Cape 1984). Variation in wind speed between these types may also affect precipitation acidity (Fowler and Cape 1984), as may other characteristics associated with these weather types. On the basis of the trajectory analyses, it appears to be well worthwhile further exploring the relationships between synoptic classifications and wet acidic deposition in Scotland.

Figure 7.1 shows the relationship between Lamb Weather Type, percentage of total H^+ deposition and percentage of precipitation amount at Eskdalemuir, Scotland. Only LWTs contributing more than 7.5 per cent of annual H^+ deposition and more than 7.5 per cent of annual precipitation have been included. The H^+ content of annual precipitation at this station is dominated by C-type and W-type events. Over the five-year period, rainfall occurring on C-type days accounted for about 30 per cent of the total deposition and around 25 per cent of the total precipitation. The respective values for W-type rainfall events are approximately 12 and 25 per cent. Although rainfall associated with other weather types make significant contributions to the concentration or dilution of precipitation acidity in certain years (S-type and SE-type events in particular), the acid content of annual precipitation is dominated by C- and W-type associated rainfall. C-type events make the major contribution to increasing the H^+ concentrations (a large average departure in the positive y-direction from the equality line, combined with a substantial contribution to the total rainfall). W-type events make the major contribution to diluting H^+ levels (greatest departure in the negative y-direction from the equality line, together with a large contribution to the annual rainfall).

It is possible to devise a simple index of the influence of atmospheric circulation variation over the UK on H^+ deposition, by subtracting the number of W-type days from the number of C-type days $(C-W)$. This should then be related to volume-weighted H^+ concentration in precipitation (Davies *et al.* 1986). A comparison of the annual $C-W$ index, for the period 1958–64, with annual precipitation-weighted H^+ concentration (using the daily EMEP precipitation chemistry data (1978–84) and the monthly EACN precipitation chemistry data back to 1958 (Davies *et al.* 1986), shows a correlation of 0.37 (significant at the 5% level). (Davies *et al.* 1986 argue that there are grounds for excluding two years from the analysis, whereupon the value of the correlation coefficient rises to 0.6.) However, the relationship between the two variables cannot be stationary in time, because of changes in other important factors such as varying emissions. A more appropriate measure of association is, therefore, a sign test on the year-to-year

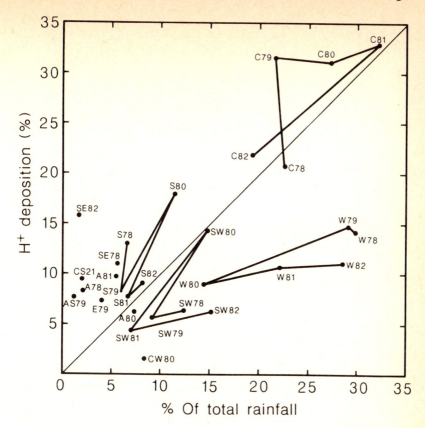

Figure 7.1 Relationship between Lamb Weather Type, percentage of total H$^+$ deposition and precipitation amount at Eskdalemuir

Note: The points represent total precipitation falling on particular weather-type days for each year. W82 represents westerly 1982 and so on. C is cyclonic. The other weather types can be identified from Lamb (1972a). For clarity, the points for the major weather type contributors have been joined.

fluctuations. In this case, the test statistic (including all years 1958–84) is significant at the 0.1% level.

The observed association between the C−W index and annual precipitation-weighted H$^+$ concentration at Eskdalemuir allows consideration of the possible implications for acidic deposition of the known long-term changes in atmospheric circulation over the UK. Figure 7.2 shows the long-term history of the C−W index. Allowing for 'anomaly' in H$^+$ concentration in the mid-1970s (Davies *et al.* 1986), the five-year volume-weighted H$^+$ concentrations at Eskdalemuir show a closer match with the C−W record than with the UK emissions of sulphur dioxide (sulphate has been the anion best correlated with precipitation acidity in Scotland; see UKRG 1983). So, besides the strong relationship on the year-to-year time-scale, discussed above, there may also be links on longer time-scales. Unfortunately, the H$^+$ record is too short to test the point statistically. However, the associations recognised thus far show that the climate dimension must be taken into

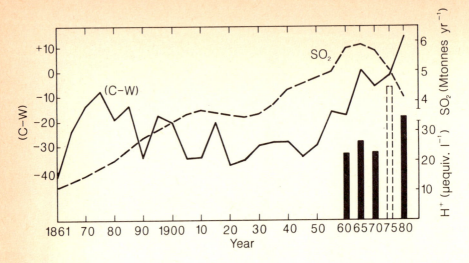

Figure 7.2   Five-year means for the C−W index

Note: SO₂ emissions estimated from published data (Davies *et al.* 1986); and five-year volume weighted H⁺ concentrations at Eskdalemuir. In the latter case, the pentade 1975–9 is represented differently to highlight the anomalous nature of the 1975–6 period, and the use of two different data sources during this interval (Davies *et al.* 1986).

account in any consideration of the acid-rain issue. The C−W index may be criticised on the grounds of over-simplicity, so it is necessary to examine the influence of specific meteorological variables.

The influence of precipitation amount on sulphate and nitrate deposition

Although sulphate has been best correlated with precipitation acidity in Scotland, the contribution of nitrate has been increasing in recent years (UKRG 1987). The mean concentration in precipitation of a chemical species is one measure of pollution but of possible greater ecological significance is the total input into an area (concentration of a species multiplied by the precipitation amount). Here we examine the relationships between both concentration and deposition of nitrate and sulphate (corrected for the sea-salt component) in precipitation at Eskdalemuir.

EACN monthly data for the period 1958–79 were used to compile precipitation-weighted annual means. On an annual basis, the nitrate deposition is controlled by nitrate concentration, whereas sulphate deposition is more strongly driven by the amount of associated precipitation (Table 7.1). These relationships are understandable in terms of what is known of the oxidation processes of NO_X and SO_2 (Farmer *et al.* 1987). The important implication of the relationships is that any variability in precipitation amount alone would produce differential changes in nitrate and sulphate deposition. This may occur irrespective of other factors such as emission mix and rates. Time series for annual precipitation, sulphate concentration and deposition, nitrate concentration and deposition (Figure 7.3) show that nitrate deposition at Eskdalemuir has increased during the period 1958–82, but there is no discernible trend in sulphate deposition. This is in spite of the fact that

Table 7.1   Correlations between annual values of nitrate and sulphate deposition (EACN monthly data), ion concentration and precipitation amount[1]

	Ion concentration	Precipitation amount
Nitrate deposition	0.86^2 (0.80^2)	-0.16 (0.11)
Sulphate deposition	0.37^4 (0.47^2)	0.74^2 (0.77^2)

Notes: [1] The correlation values in brackets relate to annual data which were linearly detrended to minimise the effects of autocorrelation on significance levels
[2] $p < 0.001$
[3] $p < 0.05$
[4] $p < 0.10$

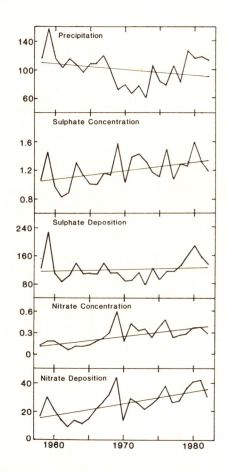

Figure 7.3   Eskdalemuir annual time series of mean monthly precipitation (mm), mean monthly excess sulphate concentrations (mg l^{-1}), mean monthly excess sulphate deposition (mg m^{-2}), mean monthly nitrate concentration (mg l^{-1}), mean monthly nitrate deposition (mg m^{-2}).

Note: Mean monthly values are plotted to compensate for years with some missing data (Farmer *et al.* 1987).

sulphate concentrations have increased by around 25 per cent during the period of study. The trends cannot be readily explained by the emission rates alone (Farmer *et al.* 1987). The rise in nitrate deposition is clearly driven by increasing nitrate concentrations, while the sulphate deposition is 'suppressed' by the constant or slightly decreasing precipitation totals. There is an element of uncertainty to the argument when linear trends are fitted to such curves, but the period 1967–74 represents a clear example. Sulphate concentrations exhibit their major peak (above the trend line) during this period, and yet sulphate deposition exhibits a trough, because of lower than average precipitation.

Monthly wet deposition and atmospheric circulation patterns

Changes in the frequency of different types of trajectory and variations in precipitation amount, evident over times-scales of years to decades, are related to changes in the regional-scale circulation patterns. We have already explored the influence of changing climate on the long-term variations in the composition of rainfall in Scotland. Over the last 20 years, the greatest interest in acidic deposition in Europe has focused on southern Scandinavia. Over the period 1970–9 of the EACN record, observations of sulphate (corrected for sea-salt component) concentrations in monthly precipitation collected at the stations shown in Figure 7.4 were examined in order to identify 'high'-concentration and 'low'-concentration months in the area. The ten months of 'highest' concentration and the ten months of 'lowest' concentration being thus identified, the monthly sea-level pressure distributions (using the UK Meteorological Office gridded data set, $5° \times 10°$ longitude to 65 °N and a reducing grid to 90 °N) were expressed as anomalies from the long-term mean for each month. A map was then produced of the mean anomaly sea-level pressure fields for (i) the ten 'high' sulphate concentration months and (ii) the ten 'low' sulphate concentration months. For the anomaly data, a *t*-test of the mean anomaly was carried out at each grid point. Where the *t*-test is significant at the 5% level, it is flagged on the mean anomaly map with an asterisk. The question of field significance is complex and is not easy to resolve (Preisendorfer and Barnett 1983; Wigley and Santer 1988), but even without 'significant' points many of the maps produced are physically sensible with spatially coherent and discernible regions. The statistically significant points can be seen as a way of highlighting the 'centres of action'.

Figure 7.5 shows the mean pressure anomaly map and the mean actual sea-level pressure map for the ten high sulphate concentration months. The individual grid-point values and the asterisk-flagging have been omitted for clarity. The anomaly pattern indicates higher than normal pressure over Scandinavia, and lower than normal pressure over much of western Europe. Enhanced flow from the south and east is suggested. Wet acidic episodes in Scandinavia are known to be associated with outflow from anticyclones over eastern Europe and subsequent removal of the pollution by frontal precipitation of Atlantic depressions (Smith and Hunt 1978). There are large sulphur sources in eastern Europe. This interpretation is confirmed by the mean actual sea-level pressure distribution, which indicates meridional flow over Scandinavia, with a strong ridge of high pressure extending from the east. It is not suggested that all precipitation events with high sulphate concentrations were associated with back-trajectories from the east, but the zonal flow was much weaker than normal.

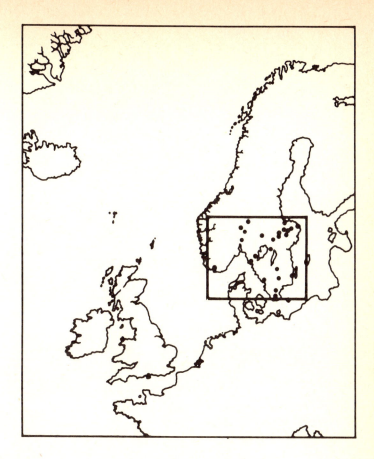

Figure 7.4 Location of observation stations used to identify 'high' and 'low' excess sulphate concentration months in southern Scandinavia

The low sulphate concentration months were characterised by a reverse anomaly pattern (Figure 7.6), with a negative anomaly over Scandinavia and a positive anomaly over western Europe. This would have produced enhanced flow from the north and west (no pollution sources), and this is represented on the mean actual pressure distribution as a strong zonal flow over southern Scandinavia. Over this ensemble of months, there was clearly a tendency for transport to southern Scandinavia to originate over the maritime regions.

This example of the use of pressure and pressure anomaly composite maps has demonstrated that this is a useful technique to gain insight into the influence of regional-scale circulation patterns of ionic concentrations in precipitation. It is known that there have been marked changes in various circulation indices in the North Atlantic this century (Lamb 1972b), and so it is perfectly possible that the chemical composition of precipitation over Europe has varied, on this time-scale, in response to those changes.

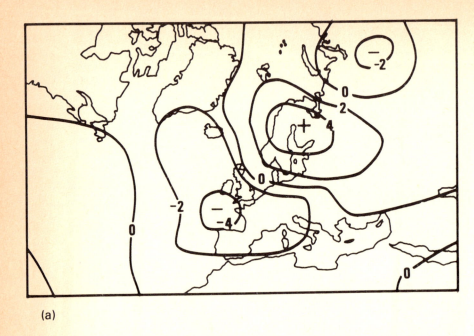

(a)

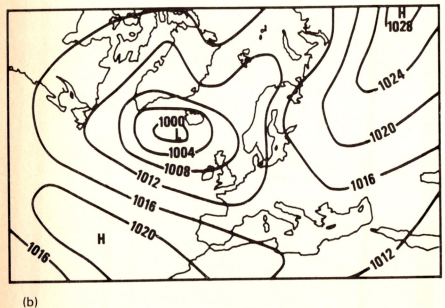

(b)

Figure 7.5 (a) Composite pressure anomaly (mb) pattern for 'high' excess sulphate concentration months in southern Scandinavia. (b) Composite actual sea-level pressure distribution for the same months

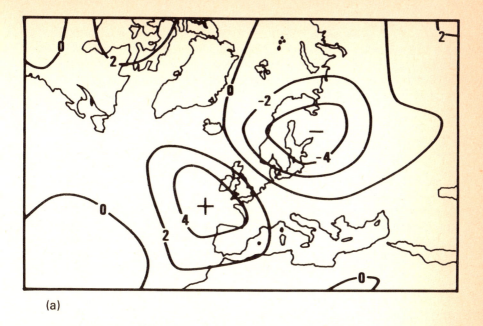

(a)

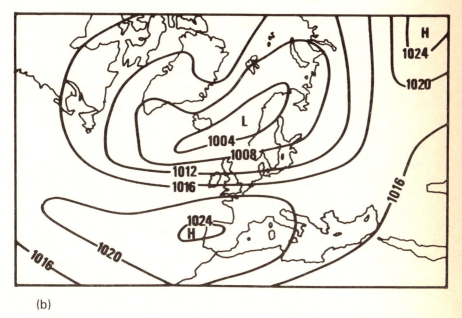

(b)

Figure 7.6 (a) Composite pressure anomaly (mb) pattern for 'low' excess sulphate concentration months in southern Scandinavia. (b) composite actual sea-level pressure distribution for the same months

Conclusions

There has been much recent interest in the question of 'non-linearity' relating to acidic deposition on a regional scale. It is possible that chemical mechanisms in the atmosphere may produce non-proportional relationships between emissions and the deposition of acidic species. In this paper we have considered wet deposition. There are significant links between the wet deposition of hydrogen ions at a station in southern Scotland and trends over years in the frequency of weather types affecting the UK. When precipitation amount only is considered, it is demonstrated that variations from year to year can alter the nitrate to excess sulphate ratio in wet deposition in Scotland (the relationship need not necessarily hold elsewhere). This may be related to the oxidation of NO_X in the gas phase, and the production of sulphate in liquid phase reactions (Calvert *et al.* 1985). Given this, it is to be expected that the amount of sulphate deposited will tend to correlate with precipitation amount, whilst the amount of nitrate deposited is likely to be independent of precipitation amount. An alternative explanation could be systematic changes over the years in the frequency of precipitation association with particular air masses with their own characteristic sulphate to nitrate concentration ratios (Davies *et al.* 1986). These links have implications for any changing attribution of precipitation acidity.

Trends in weather-type frequency, trajectory frequency and precipitation amount are produced by changes in regional-scale atmospheric circulation. Groups of months which exhibit 'high' or 'low' concentrations of non-sea-salt sulphate in precipitation may be explained, at least in part, by the monthly sea-level pressure distribution which may predispose a receptor region (in this case, southern Scandinavia) to precipitation events associated with trajectories either from source regions or from regions with few sources of pollution. These relationships allow the possibility of considering the contribution of changing atmospheric circulations over Europe to the 'non-linearity' of emission–deposition relationships. This may become important if there are significant trends in the character of atmospheric circulation on time-scales of years to decades.

Acknowledgements

The work presented in this paper was supported by the Department of the Environment and the Commission of the European Community. The contribution of G.M. Glover is from work carried out at the Central Electricity Research Laboratories and is published by kind permission of the Central Electricity Generating Board.

References

Calvert, J.G., Lazrus, A., Kok, G.L., Heikes, B.G., Walega, J.G., Lind, J., Cantrell, C.A. 1985. 'Chemical mechanisms of acid generation in the troposphere', *Nature* 317, 27–35.
Davies, T.D., Kelly, P.M., Brimblecombe, P., Farmer, G., Barthelmie, R.J. 1986. 'Acidity of Scottish rainfall influenced by climatic change', *Nature* 322, 359–61.
Davies, T.D., Brimblecombe, P., Tranter, M., Abrahams, P.W., Blackwood, I.L. 1988. 'Chemical composition of snow in the remote Scottish Highlands' in M.H. Unsworth and D. Fowler (eds), *Deposition Processes in Mountains*, NATO.

Farmer, G., Barthelmie, R., Davies, T.D., Brimblecombe, P., Kelly, P.M. 1987. 'Relationships between concentration and deposition of nitrate and sulphate in precipitation', *Nature* 328, 787–9.

Fowler, D., Cape, J.N. 1984. 'On the episodic nature of wet deposited sulphate and acidity', *Atmos. Environ.* 18,9. 1859–66.

Jones, P.D., Kelly, P.M. 1982. 'Principal component analysis of the Lamb Catalogue of Daily Weather Types: Part 1, Annual Frequencies', *J. Climatol.* 2, 147–57.

Lamb, H.H. 1972a. 'British Isles weather types and a register of the daily sequence of circulation patterns, 1861–1971', *Geophys. Mem.* 116, London.

Lamb, H.H. 1972b. *Climate Present, Past and Future. Vol. 1 Fundamentals and Climate Now*, London, Methuen.

Preisendorfer, R.W., Barnett, T.P. 1983. 'Numerical model-reality intercomparison tests using small-sample statistics', *Atmos. Sciences* 40, 1884–96.

Rodhe, H., Granat, L. 1984. 'An evaluation of sulphate in European precipitation, 1955–82', *Atmos. Environ.* 18,12, 2627–39.

Schaug, J., Dovland, H., Skjelmoen, J.E. (various years) *ECE Cooperative Programme for Monitoring and Evaluation of the Long-range Transmission of Air Pollution in Europe*, Data Reports, October 1979–September 1982, Lillestrom, Chemical Co-ordinating Centre, Norwegian Institute for Air Research.

Smith, F.B., Hunt, R.D., 1978. 'Meteorological aspects of the transport of pollution over long distances', *Atmos. Environ.* 12, 461–77.

UK Review Group on Acid Rain 1983. *Acid Deposition in the UK*, Stevenage, Warren Spring Laboratory.

UK Review Group on Acid Rain 1987. *Acid Deposition in the UK 1981–1985, A second Report*, Stevenage, Warren Spring Laboratory.

Wigley, T.M.L., Santer, B.D. 1988. 'Validation of general circulation climate models' in M.E. Schlesinger (ed.), *Physically-based Modelling and Simulation of Climate and Climatic Change*, Dordrecht, D. Reidel.

Chapter 8

Intra-seasonal fluctuations of mid-tropospheric circulation above the eastern Mediterranean

J. Jacobeit University of Augsburg,
Federal Republic of Germany

Introduction

Recent investigations by Sharon and others have revealed some intra-seasonal fluctuations that are masked by monthly averaging, but appear in time-consistency if several years of submonthly ten-day periods are averaged together. Rainfall amounts of 17 Israeli stations during 1950–75 (Lomas *et al.* 1976), the frequency of stormy days in Israel during nine winter seasons between 1964 and 1976, the frequency of Cyprus lows above Israel between 1965 and 1971 and the position of upper troughs and the polar front jet stream above the Mediterranean area for selected ten-day periods averaged over 1974–83 (Sharon and Ronberg 1986), all show a regular sequence of more and of less active periods between October and April which last for 10–50 days and differ, for example, in the frequency of stormy days by a factor of two. Figure 8.1 shows mean monthly and mean ten-day rainfall amounts for the months of October–April during 1966–76 for six northern or central Israeli stations. In contrast to the smooth regular course of monthly means the ten-day periods reveal three major breaks in the seasonal progression around early January, mid-February and early April with more active periods in between, before and after them. In order to examine whether this intra-seasonal fluctuation originates in behaviour of the general circulation an independent data set has been used (Jacobeit 1985) comprising daily heights of the 500 hPa level and numerically evaluated elements of current within the Atlantic–European region (20–70 °N, 57.5 °W–62.5 °E) for the decade 1966–76 which only overlaps in part the other periods mentioned above.

Elements of current above Israel

Cyclonic elements of the 500 hPa circulation above the Mediterranean area will be distinguished according to their amplitude and their range of influence. Troughs are regarded as waves with large meridional extent covering both areas north and south of 45 °N, other cyclonic elements as cells or waves of low amplitude influencing the lower mid-latitudes south of 45 °N. Figure 8.2 shows the relative frequencies of troughs, other cyclonic elements and anticyclonic elements above Israel from October to April, averaged for the decade 1966–76 and subdivided into ten-day periods. Most conspicuously, the troughs' curve shows exactly the same breaks in early January, mid-February and early April, with more active periods before, between and after. Only the trough minimum at mid-November does not correspond with the rising rainfall curves, but at this time the frequency of cyclonic elements show various stages. After the general decline towards the end of the year, which runs contrary to the general increase in trough frequency and rainfall, a more or less parallel progression with trough frequency prevails until mid-March. After that the opposite development predominates again. This alternation might be explained by the dominance of high-amplitude trough activity (enhanced or weakened) during the mid-winter season (both as direct trough influence and as induced cut-off effectiveness), whereas the transitional periods allow for greater representation of low-amplitude elements when trough activity decreases. Frequency variations of the anticyclonic elements are less distinct but some developments inverse to the troughs' curve may be seen as a general tendency.

Eastern Mediterranean troughs

Figure 8.3 shows the average ten-day frequency of upper trough axes within the extended range of the eastern Mediterranean (20–42.5 °E) as far as trough amplitudes extend across 45 °N. Not all of these troughs do reach the Israel region directly (mainly because of lower amplitudes or lesser zonal extents), but frequencies in Figure 8.3 and Figure 8.2 show general accordance. According to Sharon and Ronberg (1986), Figure 8.3 allows the definition of several subseasons of variable duration the relative trough frequencies of which are given in Table 8.1. The mean values of the more and of the less active subseasons differ by a factor close to two with greater differences until January–February and smaller ones afterwards. These subseasons well agree with those defined by Ronberg on the different basis of frequency of stormy days in Israel; minor deviations which do not exceed a single ten-day period may arise from the different variables that are considered. Thus the time-consistency still improves with respect to the submonthly frequencies of Cyprus lows which are more directly linked to the eastern Mediterranean upper trough activity. On the other hand the extrema do not have equal significance. The maxima tend to have higher interannual variability concerning both their absolute level and their temporal occurrence, the minima imply fewer deviations. Individual departures must be more seldom or of insignificant dimension to maintain the overall attribute of minimal occurrence; moreover, individual less active periods take place at time-spans mostly covering a central subperiod which thus shows a striking minimum character in the mean. This implies that the interannual variations of trough frequency are concentrating on the active subseasons whereas the culminating break phases are not covered on more than one occasion out of ten by an active spell of striking character.

Mediterranean trough axis distributions

Figure 8.4 presents the frequencies of upper trough axes within the subseasons of Table 8.1 for the whole Mediterranean area (5 °W to 40 °E) subdivided into six bands of 7.5° longitudinal extent each. Quite evidently, the variations are not restricted to the eastern long-wave area, but also occur in the region upstream, thus confirming intraseasonal fluctuations of the mean trough axis position (Sharon and Ronberg 1986). Subseason 2 (more active) differs from subseason 1 (more settled) not only in its generally elevated level of frequency, but also in the trough axis

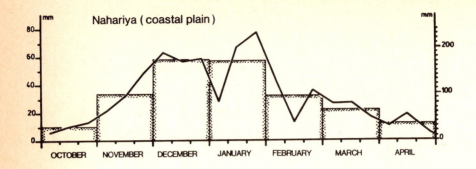

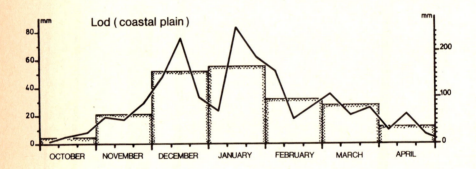

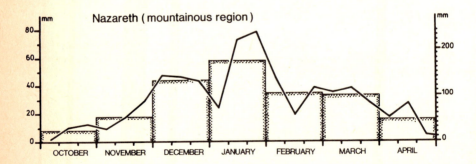

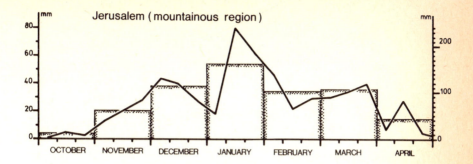

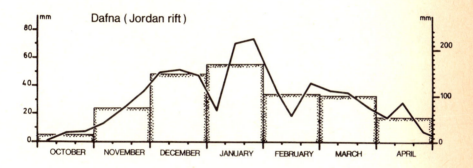

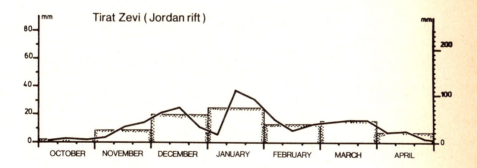

Figure 8.1 Mean monthly[1] and mean submonthly ten-day rainfall amounts[2] for the months of October–April, 1966–76, at six Israeli stations

Note: [1] Stippled columns with scale to the right
　　　 [2] Solid line with scale to the left.

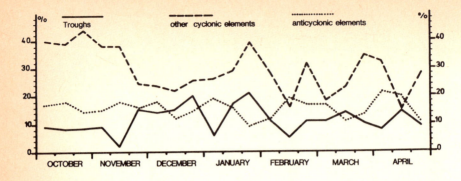

Key:

– – – –   other cyclonic elements
. . . . . .   anticyclonic elements
_____   relative frequencies of upper troughs

Figure 8.2   Relative frequencies of upper troughs, other cyclonic elements and anticyclonic elements of the 500 hPa level above Israel for ten-day periods in the months of October–April, 1966–76

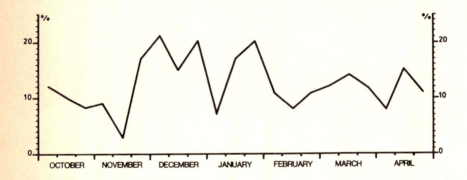

Figure 8.3   Relative frequencies of upper troughs (500 hPa level) in the eastern Mediterranean (trough axis between 20 and 42.5 °E) for ten-day periods in the months of October–April, 1966–76)

maximum shifted eastward. The break phase of early January exhibits a striking maximum two bands further west (10–17.5 °E), whereas the following active period again reveals an eastward-shifted maximum with lower frequencies, however, in the western half contrary to the former active period. The following more settled subseason (February) shows broadly distributed (2.5–25 °E) medium-sized frequencies to the west, whereas the active period during March still develops its absolute maximum in the west (2.5–10 °E) in spite of the enhanced activity to the east. Subseason 7 (more settled) again reveals a central maximum (10–17.5 °E) which is increasing frequencies to the east. This indicates that shifting of the mean trough axis and some other factors operate: a higher proportion of mobile troughs manifests itself in a more balanced trough distribution among the various longitudinal bands (both at low (subseasons 1 and 5) and at high absolute level

Table 8.1 Relative frequency (%) of upper troughs (500 hPa level) in the eastern Mediterranean (trough axis between 20°E and 42.5°E) for various subseasons between October and April, 1966–76

Subseason	Relative frequency (%)
1 (1.10.–20.11.)	8.4
2 (21.11.–31.12.)	18.3
3 (1. 1.–10. 1.)	7.0
4 (11. 1.–31. 1.)	19.0
5 (1. 2.–28. 2.)	9.9
6 (1. 3.–31. 3.)	12.6
7 (1. 4.–10. 4.)	8.0
8 (11. 4.–30. 4.)	13.0
Average of the less active subseasons:	8.3
Average of the more active subseasons:	15.7

(subseasons 2 and 6)). More accentuated distribution patterns with preferred ranges of trough activity, on the other hand, indicate a higher proportion of quasi-stationary troughs (during subseasons 3 and 7 with the mean trough axis above the central Mediterranean, during subseason 4 with the maximum in the long-wave range). Finally, subseason 8 reveals a broad and at the same time striking maximum which indicates an elongated centre of action or more frequent successions of initially stationary and subsequently mobilised troughs.

The shifting of the mean trough axis may be seen more clearly from the mean positions of the trough axis maxima during consecutive ten-day periods (Figure 8.5). Autumn is characterised by a continued westward shifting which is reflected in the first subseason of low activity above the eastern Mediterranean. At the beginning of the following active period the mean trough axis suddenly shifts to the east (28 °E) from where it gradually recedes until a steeper regression to the west (12.25 °E) takes place just at the break phase of early January. Subsequently the mean trough axis shifts to the east again (more active period of January) before another retrogression sets in that culminates in mid-February (second break phase). After that, more easterly regions again are governed by the trough axis maximum (more active period of March), whereas the following regression already culminates at the end of March (break phase not before early April). It is at this time that the other cyclonic elements above Israel increase in frequency (Figure 8.2) and still maintain a higher level of submonthly rainfall. The rainfall restoration after the third break, despite a continued central trough axis maximum (12.25 °E), is caused by the highly broadened range of above-average trough frequency during subseason 8 (Figure 8.4).

Circulations upstream

In order to integrate the Mediterranean trough axis distribution patterns into the general conditions of circulation, various parameters of the 500 hPa circulation above the crucial upstream region of the North Atlantic have been calculated as mean values for each ten-day period for the months of October–April during the

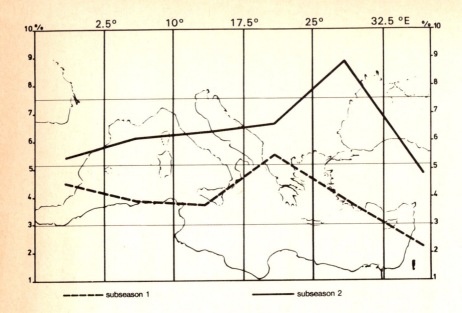

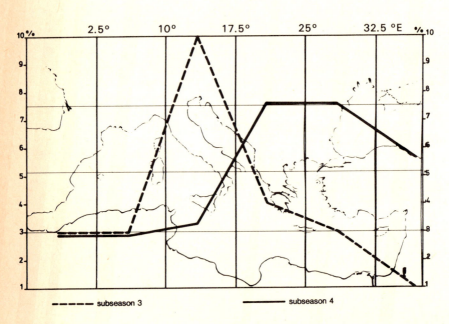

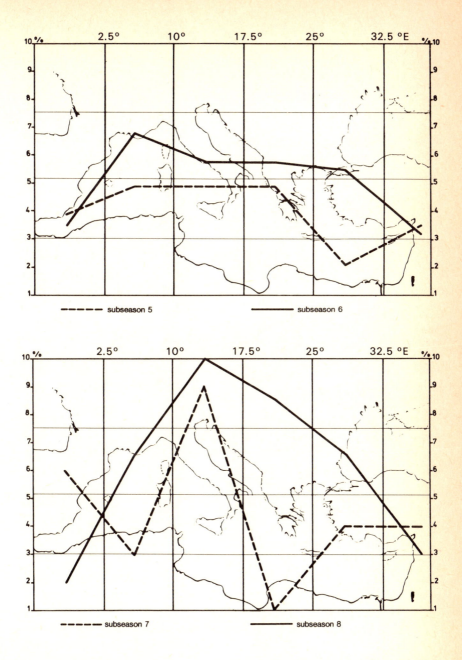

Figure 8.4 Relative frequencies of upper trough axes (500 hPa level) in six longitudinal strips during subseasons defined in Table 8.1

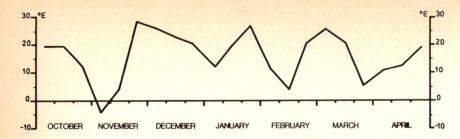

Figure 8.5 Mean longitudinal position of the upper trough axis maxima (500 hPa level) in the Mediterranean area for ten-day periods in the months of October–April, 1966–76

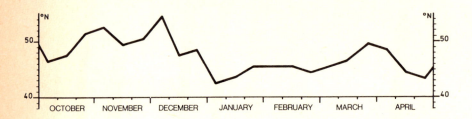

Figure 8.6 Mean latitudinal position of the main branch of the westerlies (500 hPa level) above the Atlantic for ten-day periods in the months of October–April, 1966–76

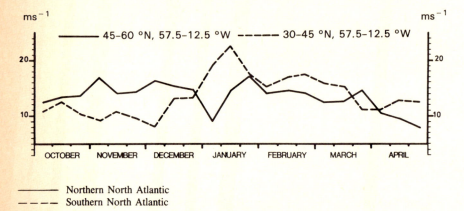

———— Northern North Atlantic
– – – – Southern North Atlantic

Figure 8.7 Mean zonal index (ms^{-1}) at the 500 hPa level above the northern and the southern North Atlantic for ten-day periods in the months of October–April, 1966–76

period 1966–76: the mean latitudinal position of the main branch of the westerlies above the North Atlantic (Figure 8.6); the mean zonal indices (Figure 8.7) for two latitudinal sections of the Atlantic (45–60 °N and 30–45 °N, each between 57.5 and 12.5 °W); a measure for the vertical components of the relative vorticity

$$\Sigma = \frac{\partial v}{\partial x} - \frac{\partial u}{\partial y}$$

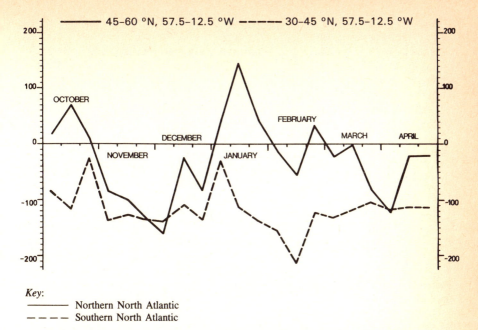

Key:
——————— Northern North Atlantic
— — — — Southern North Atlantic

Figure 8.8   Mean measure of the relative vorticity at the 500 hPa level above the northern and the southern North Atlantic for ten-day periods in the months of October–April 1966–76

(Figure 8.8) which results from approximating geopotential gradients depending on latitude (see Jacobeit 1985: 46).

Apart from minor fluctuations, some intrinsic tendencies appear. The first more active period above the eastern Mediterranean is still not accompanied by drastic changes in the upstream circulation and thus originates from the seasonally changed meridional heat balance above predominantly continental regions or from reductions of the hemispherical wave number. Drastic changes, however, take place in early January: the main branch of the Atlantic westerlies shifts southward by 6° of latitude to its most southerly mean position (42.5 °N), the zonal indices correspondingly develop inversely (steep decline in the northern, strong increase in the southern North Atlantic), the northern relative vorticity assumes cyclonic sign, the southern one reaches its relatively highest (least anticyclonic) value. This sudden transformation to high-winter conditions above the Atlantic results in forced reactions of the downstream circulation: in some northward shift of the eastern Atlantic anticyclonic centre of action, in the downstream cyclonic turn above the central Mediterranean and in the break phase (concerning trough frequency and rainfall) above the eastern Mediterranean area. Large-scale current types with anticyclonic control above the eastern Atlantic or western Europe and central Mediterranean troughs, low-amplitude cyclonic waves or cut-off cells now occur with roughly twice the mean frequency (Jacobeit 1985). At the same time trough frequency above the western Atlantic is raised considerably in connection with the southward advancing main branch of the westerlies. Thus the relative vorticity above the southern North Atlantic (already increased by the northward shift of the anticyclonic centre) assumes its highest value. The trough axis maximum above the central Mediterranean, however, turns out to be highly variable since it is often replaced by various cut-off stages. In any case, however, the eastern Mediterranean is seldom reached by this cyclonic activity and experiences its well-known break phase.

In the following weeks this pattern shifts to the east (passing through a pronounced cyclonic maximum above the northern North Atlantic, and eastward drifting of the Mediterranean trough axis maximum) with the main branch of the Atlantic westerlies only receding a little and persisting in relatively low latitudes until March. The second break phase in the eastern Mediterranean which arises from another westward shifting of the mean trough axis, is characterised by steeply declining Atlantic vorticities into moderately negative values above the northern and extremely anticyclonic values above the southern part. This coincides with roughly double the frequency of the normal occurrence of large-scale current types with anticyclonic control above the eastern Atlantic or western Europe and central Mediterranean troughs during mid-February (Jacobeit 1985). In contrast to early January, however, there is no distinct shift in the latitudinal position of the Atlantic main branch of the westerlies; moreover, the intensified anticyclonic control seems to be widened rather than shifted considering the further decreasing vorticity above the southern North Atlantic.

The following active period in the eastern Mediterranean (mean trough axis moving eastward) is characterised by dissolutions of this anticyclonic anomaly (normalising relative vorticities above the Atlantic, a more balanced trough distribution pattern above the Mediterranean area with higher proportions of mobile troughs). At the end of March the mean trough axis already shifts westward with the Atlantic main branch of the westerlies again taking a more northern path and zonal indices which are inverted in relation to each other. The continuation of more active conditions above Israel is the obvious result of elevated frequencies of Mediterranean cyclonic waves and cells.

The third break phase in the eastern Mediterranean in early April first shows little differences compared to the end of March (a rather westerly mean trough axis position, distinctly anticyclonic conditions above the northern North Atlantic, elevated frequency of Mediterranean cyclonic elements). On the other hand, the zonal index above the northern North Atlantic is decreasing at a scarcely changed position of the westerlies' main branch, and the Atlantic relative vorticity is more anticyclonic than in mid-February above the northern part, less anticyclonic above the southern part. This hints at anticyclonic shiftings of some other kind: large-scale current types with dominating troughs extending from north-western Europe to the region of the Canary Islands occur with more than twice the average frequency (Jacobeit 1985), thus influencing both the Atlantic distribution of the relative vorticity and the rather westerly position of the mean trough axis. This frequency peak might be linked to the polar anticyclone which regularly develops in springtime. In most cases this configuration is coupled with downstream waves of low amplitude and low intensity, so that, above the eastern Mediterranean, the observed break phase may well coincide with high frequencies of such waves.

The following period of mid- to late April shows another spell of southward advancing Atlantic westerlies with opposite changes of the zonal indices; in contrast to early January, however, there is no concentration of quasi-stationary troughs on limited ranges in the central Mediterranean, but a much more broadened trough axis maximum that causes reactivated rainfall conditions above Israel. This pattern which indicates a widened range of trough occurrence or an increased tendency towards subsequent mobility might result from the less effective forcing power that similar developments attain in springtime compared to the more vigorous high-winter situation.

Conclusions

The main restriction of the present study is the shortness of its data-covered period. Results and arguments, however, are considerably strengthened by the fact that similar results have been obtained from different data sets and are related to some periods that are at least in part independent of our own decade. Thus the main results may be summarised as follows. First, the frequency of the upper troughs in the eastern Mediterranean area shows a regular succession of periods of variable duration (10–50 days) with alternating higher or lower values differing by a factor of about two (with greater differences until January–February and subsequently smaller ones). Second, this intra-seasonal fluctuation coincides well with analogous fluctuations in some independent data sets concerning both the timing of phases and the degree of their differences (Sharon and Ronberg 1986). These corresponding fluctuations are reflected in submonthly rainfall amounts above Israel. Third, while the more active periods show a rather great variance concerning inter-annual variations of absolute level and peak timing, three culminating break phases of relatively high internal consistency become apparent at recurring periods around early January, mid-February and early April. Finally, these break phases seem to be linked to different stages of the seasonal development of the general circulation which may be recognised in terms of Mediterranean trough axis positions and some characterising parameters of the Atlantic circulation upstream.

References

Jacobeit, J. 1985. *Die Analyse großräumiger Strömungsverhältnisse als Grundlage von Niederschlagsdifferenzierungen im Mittelmeerraum*, Würzburger Geographische Arbeiten, 63.

Lomas, J., Zemel, Z., Shashua, J. 1976. 'Rainfall probabilities in Israel as a basis for agricultural planning', Agrometeorological report of the Israel meteorological service.

Sharon, D., Ronberg, B. 1986. 'Time-consistent fluctuations of weather patterns during the winter in Israel' in Instituto Juan Sebastian Elcano (eds), *Cambios Recientes en Climas Mediterraneos*, Madrid, 97–108.

Chapter 9

Intra-annual weather fluctuations during the rainy season in Israel

D. Sharon and B. Ronberg   Hebrew University of Jerusalem, Israel

Introduction

The winter season of any given year in the mid-latitudes consists of alternating spells of stormy and more settled weather, resulting from the migration of successive weather systems over the area. The timing of respective spells appears to vary at random from year to year. However, in a few regions in Europe and North America so-called 'climatic singularities' have been found to recur around some typically fixed calendar dates (Barry and Perry 1973). In the eastern Mediterranean very little was known hitherto on any sort of regularity in the timing of alternating weather systems and on climatic singularities arising from them. Hence, the 'average' seasonal pattern of change overriding the above irregular, short-term weather alternations has been mostly perceived of as a smooth, bell-shaped annual cycle consisting of a gradual increase in the frequency (and the intensity) of stormy weather during the autumn, a subsequent peaking around mid-winter and an ultimate decrease or phasing-out in the spring. This notion is well supported by empirical data taken on a monthly basis, such as of precipitation totals, temperature, etc. (Figure 9.1a).

Recently, new evidence has been presented in a synoptic-climatological study by Ronberg (1984) of a regular intra-seasonal fluctuation in the frequency of stormy weather that becomes apparent in long-range averages of submonthly ten-day periods, i.e. 1–10, 11–20 and 21–30 of each month (Figure 9.1b). Unfortunately, data of this type are less frequently published than monthly ones. But even the limited extent of data presently available clearly represents a tendency for greater frequencies of extreme weather during a number of separate periods (subseasons) occurring around fixed calendar days. In between these subseasons there are intermediate periods characterised by more settled conditions, or at least by a consistently lower frequency of stormy weather. Approximate dates defining the subseasons as identified by Ronberg (1984) are given in Table 9.1, which shows the frequency of stormy days averaged separately over each subseason. Results were based on a detailed study of nine complete winter seasons, from October to April, between 1964 and 1976. In the original study the above intra-seasonal fluctuation was identified in a great number of basic meteorological and synoptic variables both at the surface and at upper air levels, especially variables associated with the wind regime, cloudiness and storm phenomena (sand- and thunderstorms, hail, etc.).

Apart from the general climatological significance of the above phenomenon it may also be useful from an application point of view, because the realisation of a fixed pattern in individual years could add a new dimension to the rational planning and continual decision-making on rainfall-dependent activities such as the

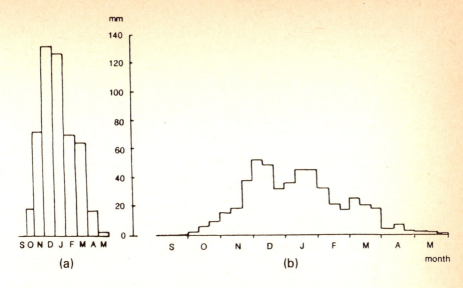

Figure 9.1 The average seasonal distribution of rainfall totals at Hafetz-Hayim (Southern Coastal Plain), 1946/7–1979/80: (a) monthly averages; (b) ten-day averages

Source: Stiefel (1981).

Table 9.1 Average percentage of stormy and of settled days during subseasons defined by Ronberg (1984), 1964–76

Subseason	Stormy days	Settled days	Other days	Stormy/ Settled
1.10–20.11[1]	7	33	60	0.21
21.11–20.12[2]	18	27	55	0.67
21.12–10.01[1]	8	37	55	0.22
11.01–10.02[2]	16	31	53	0.52
11.02–10.03[1]	7	36	57	0.19
11.03–31.03[2]	15	29	56	0.52
1.04–10.04[1]	8	32	60	0.25
11.04–30.04[2]	9	28	63	0.32
'Active' subseason average	15.0	28.8		0.52
'Settled' subseason average	7.3	34.5		0.21

1. Intermediate settled periods
2. 'Active' subseasons

management of the national water carrier in Israel, or more specifically the management of the major reservoir at Lake Tiberias and the routing of water to recharge underground storage; planning the timetable for major agricultural activities on the national level and on the level of the individual farmer; planning the timetable for construction works (roads, housing), winter and spring tourism, etc.

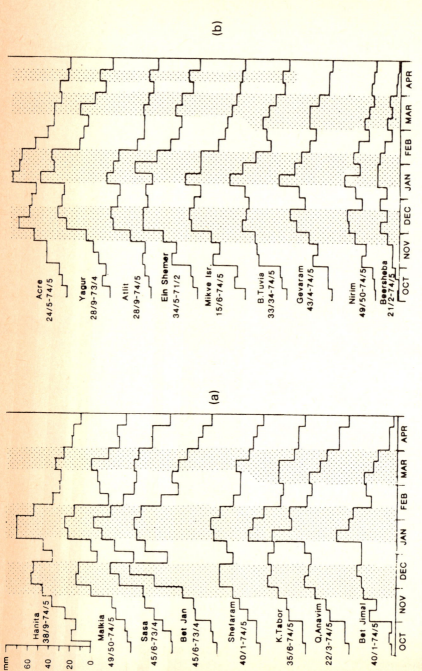

Figure 9.2 The seasonal distribution of ten-day rainfall totals (long-term averages corrected for the actual number of days at the end of each month): (a) in the mountainous part; (b) in the Coastal Plain

Source: Lomas *et al.* (1976); stippling shows active periods defined in Table 9.1.

Rainfall fluctuations

Although regular intra-seasonal fluctuations have not been noticed as such until now, they can be easily identified also in previously published data, such as on rainfall. Thus, data are available on 10-day averages of rainfall for about 20 stations, most of them with records from 30–50 years starting as far back as the 1920s or 1930s (Lomas *et al.* 1976; Stiefel 1981). Plots of these data, shown in Figures 9.1 and 9.2 reflect very clearly the pattern identified by Ronberg (1984), including also the timing of transitions between respective subseasons. This refers especially to the following features:

1. the abrupt onset of rainfall, often double-phased, in November;
2. two prominent peaks in early to mid-December and in mid to late January, with an intermediate period of two to three weeks around the turn of the year with more settled weather and lower rainfall;
3. an abrupt decrease in rainfall during mid- and late February (in the central part of the country) and extending into early March in the north; and
4. two smaller late-winter peaks, one in March immediately following the February minimum, and the other in mid-April, following a short (one to two weeks) period of more settled weather at or around the beginning of April.

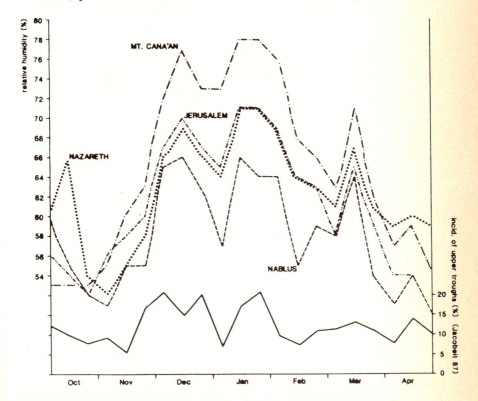

Figure 9.3 Average daily relative humidity in the mountainous part of Israel, averages for submonthly ten-day periods, 1964–79

Source: *Meteor. Serv.* (1983).

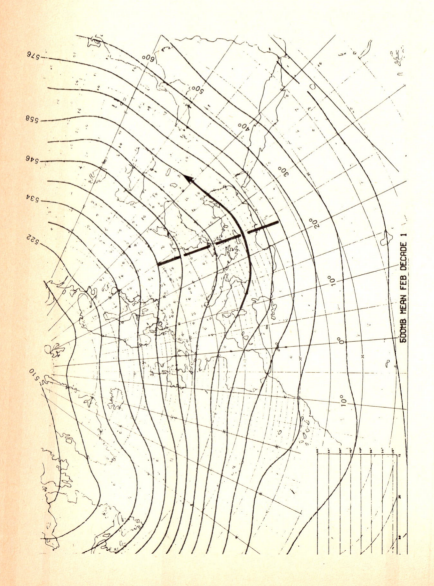

(a)

500MB MEAN FEB DECADE 1

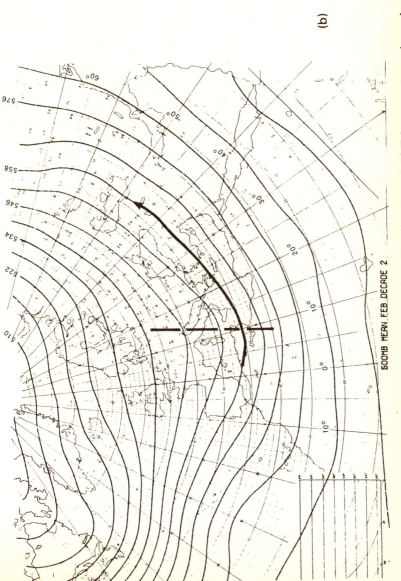

500MB MEAN FEB DECADE 2

(b)

Figure 9.4 Average 500 mb map (gpm), 1974–83, for two ten-day periods: (a) 1–10 February, representing end of an 'active' period; (b) 11–20 February, representing a settled period

Note: The approximate positions of the trough and the Polar Front Jet Stream are shown.

Source: Ronberg (1984).

Identical trends have been found in other related variables such as in cloudiness, air humidity, visibility, westerly flow, etc. An example is shown in Figure 9.3.

Fluctuations in the troposphere

Information on the immediate causes of the above fluctuation is obtained from a comparison with corresponding average seasonal changes in the synoptic domain. First, a complete series of average pressure maps (surface and upper air) has been prepared by Ronberg (1984) for all ten-day periods from October to April, averaged over ten years, 1974–83. Examples are shown in Figure 9.4. An inspection of the complete series shows a surprisingly similar seasonal fluctuation in the location of an upper air trough relative to the Levant and a north–south migration of the polar and/or subtropical jets, respectively. All four subseasons of more active weather coincide with periods of an invigorated general circulation that manifests itself in the presence of a cold upper-air trough and the positioning of the Polar Front Jet Stream over the eastern Mediterranean. The transition into more settled intermediate periods is marked by a change into less-active conditions resulting from the position of the upper-air trough east or north-eastwards, and the retreat of the jet stream northwards. The above transitions are often quite abrupt and are accompanied by significant changes in the zonal flows as in Figure 9.4.

These findings have been verified and extended in a more systematic analysis of the mid-tropospheric circulation above the Mediterranean Sea carried out by Jacobeit. The analysis is based on 500 hPa data for ten years, 1966–76. From these data the positioning of upper-air troughs above the eastern Mediterranean could be objectively identified on each day and their frequency averaged over submonthly ten-day periods as above. Results are reproduced in Figure 9.5, which shows intra-seasonal fluctuations that closely follow those presented above. In addition, Jacobeit presents results on the corresponding intra-seasonal changes in the most frequent longitudinal positioning of upper-level troughs between Gibraltar and the Levant and in the mean latitudinal position of the main branch of the westerlies above the Atlantic. Based on this, the relevant stages could be followed in the seasonal development of the general circulation, with which the main phases of the intra-seasonal fluctuations are linked. This applies especially to the three break periods of more settled atmospheric conditions during the winter that have been identified above over the Levant in early January, mid-February and early April.

Additional support for these findings has been obtained by studying the frequency and persistence of travelling cyclones over the eastern Mediterranean and the Levant (Cyprus lows). From the point of view of the rain-producing processes, the frequency of travelling cyclones passing over the study region serves as an intermediate variable between the two variables dealt with above, the frequency of upper-level troughs, on the one hand, and rainfall totals, on the other. Independent information on this has been derived here using Koplowitz's catalogue of the synoptic types over the Levant and the eastern Mediterranean (Koplowitz, 1973). The catalogue is based on a detailed objective classification of pressure field patterns obtained through Lund's method (1963), and includes a listing of types for each single day during 1965/6–1970/1. Thus it offers a unique basis, although for a limited period only, for studying the frequency and seasonality of synoptic types. Results for Cyprus lows shown in Figure 9.6 closely follow the seasonal pattern found above.

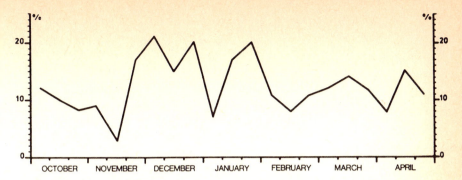

Figure 9.5 Relative frequency of upper troughs (500 hPa level) in the eastern Mediterranean between 20° and 42.5 °E, for ten-day periods in the months of October–April, 1966–76

Source: Jacobeit, Chapter 8, present volume).

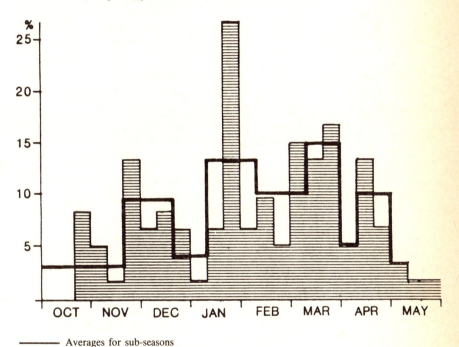

———— Averages for sub-seasons

Figure 9.6 Relative frequency of days with Cyprus low above, NW or NE of Israel during 1965/6–1970/1 for ten-day periods

Further development of results

Having established an empirical basis for the above phenomena in general, some experimentation was done to refine the time-scale of the fluctuations and to view them in a broader context in space.

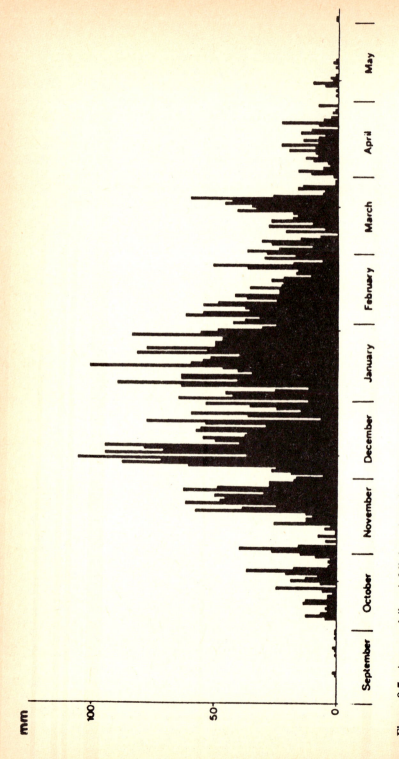

Figure 9.7 Average daily rainfall throughout the rainy season at Ein Hahoresh, 1958/9–1977/8

Source: Kutiel (1985).

The time-scale

In order to follow the timing of the fluctuations more precisely rainfall totals have been averaged over a number of years for each calendar day (Figures 9.7 and 9.8). Obviously day-to-day variations are quite irregular due to the limited extent of data available for a single calendar day. A time interval of six days has been found a good compromise, yielding both regularity and detail. Results for two stations (16–20 years) and for a cluster of four stations (ten years) are shown in Figures 9.7, 9.8 and 9.9, all from the coastal plain where localised effects on rainfall are small. Results agree well with previous findings concerning the two broad peaks in midwinter (December and January). In the autumn and spring, however, additional modulations reveal themselves. For instance, a small peak becomes apparent late in February (Figures 9.7 and 9.8) after what appears as a short lull in the wake of the January peak at the end of each month. Thus, three or four regularly spaced

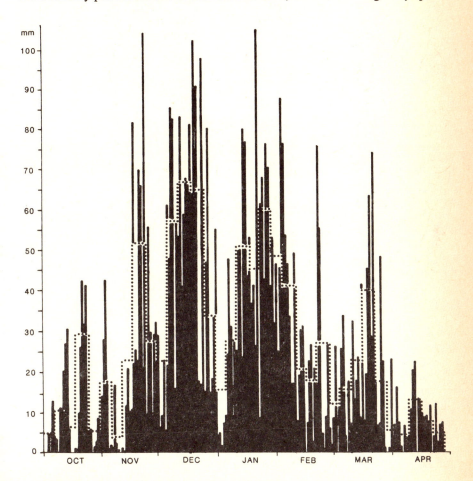

. . . . Six-day average from 1 October onwards.

Figure 9.8   Average daily rainfall throughout the rainy season at Hadera, 1963–79

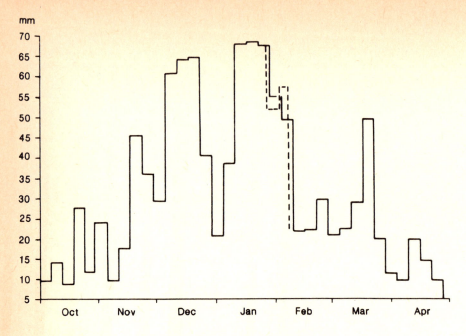

- - -   Pattern obtained from daily totals (see also Figures 9.7 and 9.8)

Figure 9.9 Rainfall totals for six-day periods from 1 October to 27 April, averaged over ten years, 1963–72, and over four stations in one area (Eyal, Hadera, Mayan Tsvi and Rupin)

peaks are clearly apparent during the spring at intervals of about 25 days. In Figure 9.7, which has been extended into May, an additional final pulsation of winter is apparent at an identical time interval. In the early autumn, the frequency of the apparent modulations is even shorter, in fact too short to be firmly defined as such.

A possible relationship to weather singularities in central Europe

As the fluctuations dealt with here were found to be linked with changes in the general circulation it is interesting to compare their pattern with that of so-called singularities in Europe. A systematic listing of singularities in western and central Europe has been presented by Lamb (1964) and is schematically shown in Figure 9.10 using the same time-scale and notation of alternating spells of weather as for the eastern Mediterranean results. It should be noted, of course, that precise agreement between weather changes in both areas should not be expected, since weather patterns resulting from changes in the general circulation need not be the same in different portions of the hemisphere (Barry and Perry 1973: 297). This follows because a given location may continue to be affected by one and the same control (such as the Azores high-pressure cell) or may lie midway between two controls, which may mask respective changes.

Although there is little correspondence in the detailed events, Figure 9.10 does reveal a striking similarity in the general pattern of the two alternating series: in

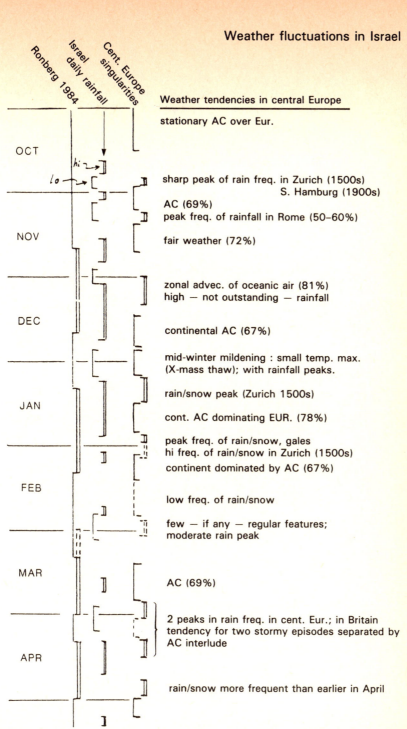

Ronberg 1984
Israel daily rainfall
Cent. Europe singularities

Weather tendencies in central Europe

stationary AC over Eur.

OCT

hi

lo

sharp peak of rain freq. in Zurich (1500s)
                               S. Hamburg (1900s)
AC (69%)
peak freq. of rainfall in Rome (50–60%)

NOV

fair weather (72%)

zonal advec. of oceanic air (81%)
high — not outstanding — rainfall

DEC

continental AC (67%)

mid-winter mildening : small temp. max.
(X-mass thaw); with rainfall peaks.

rain/snow peak (Zurich 1500s)

JAN

cont. AC dominating EUR. (78%)

peak freq. of rain/snow, gales
hi freq. of rain/snow in Zurich (1500s)
continent dominated by AC (67%)

FEB

low freq. of rain/snow

few — if any — regular features;
moderate rain peak

MAR

AC (69%)

2 peaks in rain freq. in cent. Eur.; in Britain
tendency for two stormy episodes separated by
AC interlude

APR

rain/snow more frequent than earlier in April

Figure 9.10 Intra-seasonal weather fluctuations in Israel and singularities in central
Europe prior to 1950

Source: Lamb (1964).

both, the frequency of stormy weather shows two broad peaks in December and January, with a relatively dry and warm interlude between them. Before and after that the fluctuation is characterised by narrow peaks at shorter intervals between them; in the spring the period is of 25–30 days in both series, in the autumn it seems to be even shorter.

Conclusions

Considerable evidence has been presented here on the existence of a regularly recurring intra-seasonal fluctuation in the frequency of stormy and more settled weather, respectively. Two broad peaks have been identified during December and January about 40–45 days apart, with a break of drier and warmer weather between them. From February onwards the rhythm changes into a more abrupt modulation with a shorter period of about 25 days. Peak frequencies of stormy weather occur at the end of February, after mid-March, at mid-April and possibly another one in early May. In the autumn a modulation exists too, but its pattern is more difficult to define.

The good correspondence of the results obtained here for surface variables such as rainfall, humidity etc., with synoptic variables from the middle troposphere implies that the fluctuations are associated with changes in the general hemispheric circulation. Thus, the transition from an approximate 45-day fluctuation in mid-winter to shorter ones after that is apparently associated with the shifting of the circulation from small wave numbers in the winter to greater ones in the spring.

In a broader regional context the rhythm of the above fluctuations has been found to be similar to that of the alternating weather singularities in Europe. This includes the transition from high-frequency modulations in the autumn to lower ones in mid-winter and back to high ones in early spring. Peak frequencies of respective weather types in both areas are not simultaneous, however; they differ by up to about a week, and maybe systematically so. For instance, the difference could be only in the timing of respective phases. However, since the data from both areas are related to different periods of time, the relationship between the two series can be evaluated only on the basis of information — as yet unavailable — on long-term changes of the intra-seasonal pattern.

A more detailed study of the occurrence and the timing of respective spells of weather in individual years within a given period has also yet to be undertaken, especially if practical application of results in agriculture, flood management etc., are to be considered. From the more general climatological point of view, however, it appears safe to conclude that although year-to-year variations should certainly be expected, there exists during the winter a general tendency for stormy weather to be more frequent around certain equally spaced dates, with more settled weather between respective peaks.

Acknowledgements

The basic finding reported on in this chapter has been made in a doctoral project by B. Ronberg at the Hebrew University of Jerusalem under the supervision of D. Sharon and in co-operation with the Israel Meteorological Service and its Deputy Director (Forecasting), Mr. S. Yaffe. Further analysis, reflected in Figures 9.3,

9.6, 9.8, 9.9 and 9.10 has been done jointly with S. Berkowicz, D. Lerner, K. Lubanov, O. Shani and M. Tottenauer, all graduate students at the Hebrew University.

References

Barry, R.G., Perry, A.H. 1973 *Synoptic Climatology*, London, pp. 555.

Koplowitz, R. 1973. 'An objective classification of synoptic pressure field patterns of the eastern Mediterranean basin', M.Sc. thesis, Department of Physical Geography, Hebrew University of Jerusalem, pp. 145.

Kutiel, H. 1985. 'Multimodality of the rainfall course in Israel as reflected in the distribution of dry spells', *Arch. Met. Geophys. Biocl.*, 36B, 15–27.

Lamb, H.H. 1964 *The English Climate*, London, English University Press.

Lomas, J., Zemel, Z., Shashua, J. 1976. 'Rainfall probabilities in Israel as a basis for agricultural planning.' Israel Met. Service, Agromet. Report, Bet Dagan.

Lund, I. 1963. 'Map pattern classification by statistical methods', *J. Appl. Met.* 2, 56–65.

Meteorological Service. 1983. 'Averages of temperature and relative humidity, 1964–1979'. *Meteor. Notes*, ser. A., 1(41), Bet Dagan.

Ronberg, B. 1984. 'An objective weather typing system for Israel: A synoptic climatological study', PhD thesis, The Hebrew University of Jerusalem.

Stiefel, S. 1981. *The climate of Hafetz Hayim*, Ministry of Agriculture, Rehovot (in Hebrew).

Chapter 10

Temporal changes in the spatial distribution of rainfall in the Central Coastal Plain of Israel*

Y. Goldreich   Bar-Ilan University, Israel

Introduction

There are four recognised rules which describe the spatial distribution of rainfall in Israel (see, e.g. Wolfson 1975). Three of them are universal: first, proximity to the sea increases the rainfall (except for the 3 km nearest the coast, where the influence of the convergence of the air reaching the coast is not yet manifest); second, rainfall amount increases with ground height; third, rainfall decreases in the lee of mountains. The fourth rule is a characteristic of Israel's particular geography: rainfall increases with latitude. This is because trajectories of depressions through the eastern Mediterranean usually pass north of Israel and because the influence of cold fronts, Israel's main cause of rainfall, decreases southward.

Wolfson (1975) studied the statistical relationship between the annual rainfall amount and spatial variables using multiple regression methods applied to the normals for 1931–60. He presented separate regressions for each of Israel's climatic regions. Wolfson's regression equation for the Central Coastal Plain (37 weather stations) in which the coastal city of Tel Aviv is near the centre, shows only the influence of the north/south location as significant. The regression coefficient and correlation coefficient (r) are shown in Table 10.1 (first row), where the intercept (α) is in millimetres and the independent variable N is the distance from the south (local latitude in kilometres). Residual maps based on Wolfson's regression show a positive island of residuals connecting Greater Tel Aviv with its downwind area. This finding encouraged us to add a fifth rule describing the spatial rainfall distribution, namely, the increase of rainfall in and especially downwind of urban areas (Goldreich 1981).

Recently the Israeli Ministry of Agriculture published an unofficial rainfall map of Israel for the years 1951–80. These latest normals for 34 stations of the central coastal plain were subjected to a multiple regression analysis. In these results, the independent variable, the west to east distance from the sea (D), entered the regression equation significantly, but with an unexpected positive sign implying that rainfall increases with distance from the sea (see Table 10.1). This strange anomaly can be explained in four possible ways: first, because of localised climate change; second, experimental and operational cloud-seeding which took place in the relevant area increasing the rainfall amount; third, accelerated increase in the urbanisation along the coast causing inadvertent rainfall augmentation; and fourth,

---

* The full-length version of this paper is published in *Climatic Change* (1987), 10, under the title 'Advertent/inadvertent changes in the spatial distribution of rainfall in the Central Coastal Plain of Israel'.

Table 10.1   Correlation coefficients and regression coefficients for various normals and for the whole randomised seeding experiment period

Period	No. of stations	Dependent variables	r	R	α	N	D	H	SE (D) or SE (H)
1931–60	37	All	0.59		378	0.97			
1951–80	34	All	0.79	0.88	249	1.60	3.6		
1961–75	57	All	0.85	0.85	210	2.20		0.3	0.17
1961–75	57	Seeded	0.69		105	0.69			
1961–75	57	Not seeded	0.88	0.90	106	1.50	0.3		
1961–75	57	Not seeded*	0.88	0.89	102	1.50	1.1		
1961–75	37	All	0.89	0.91	109	2.90		0.4	
1961–75	37	Not seeded*	0.89	0.90	105	3.00	1.3		0.70
1961–75	37	Seeded	0.87		26	1.30			
1961–75	37	Not seeded	0.86	0.89	89	1.60	0.3		
1961–75	37	Not seeded*	0.84	0.89	81	1.70	1.1		

* The height variable (H) was eliminated from the regression computation

Key: r = simple correlation coefficient between rainfall amount and distance from the south
   R = multiple correlation coefficient
   N = latitude (in km from south) are the regression coefficients
   D = distance from the sea (km)
   H = height (m)
   SE = standard error of the regression coefficient given in cases where it exceeds half its coefficient value.

a combination of and/or interaction among the first three factors. It is hard to ascribe any physical explanation for such an unexpected anomaly in a cold front precipitation regime to ordinary climatic fluctuations. However, the cloud-seeding effect and urban-induced changes are much more reasonable causes of such spatial changes.

The aim of this study is to differentiate between the contribution of each of these two factors that possibly influence the rainfall's spatial distribution over Israel's coastal plain. This will be achieved by examining the results of 1950–80 data regressed separately for seeded versus non-seeded days.

Cloud-seeding and urban effect on rainfall in Israel

After about 40 years of experimenting in various countries and using different methods of weather modification to increase precipitation, the results are quite disappointing. The only experiment which showed a consistent increase in rainfall and was statistically and physically sound was the Israeli one (Braham 1981; Kerr 1982). The assessment of the statistical merit of both the experiment design and modelling is based on conclusions of the Weather Modification Advisory Board (Ministry of Commerce, USA) which conducted a thorough checking of all the recent cloud-seeding experiments (Tukey *et al.* 1978).

Except for a few sporadic experiments before 1960, the first statistically designed experiment of cloud-seeding using silver iodide by planes in Israel spanned a period of six and a half years (1961–67). This randomised 'Experiment I' applied the 'crossover' technique: the seeded area was divided into two routes 'north' and 'central' (see Figure 10.1) and the seeding was carried out along one

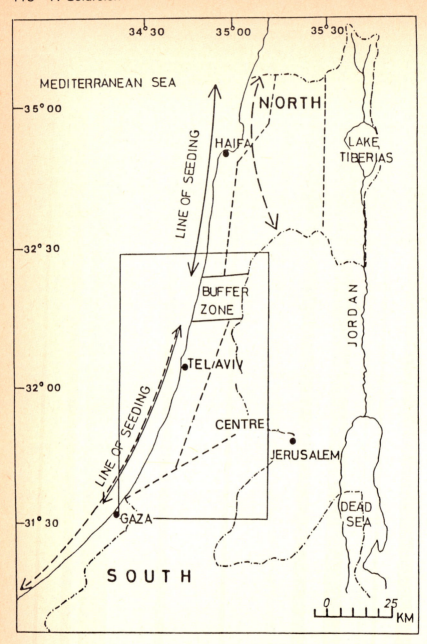

The rectangle in centre delineates the frame of Figure 10.2.

———▸ Experiment I
– – – ▸ Experiment II

Figure 10.1 The northern part of Israel with the seeding routes during Experiment I and Experiment II

Source: Gagin and Neumann (1974).

of the routes according to a predetermined calendar where the alternate non-seeded route was considered the 'control' (Gagin and Neumann 1974). A rainy day predesignated for seeding, on which, for any reason, no seeding operation took place, kept its designation as a 'seeded day'. This predetermined decision avoided any possibility of biased influence in cases where 'unsuitable' clouds were not seeded. In order to prevent seeding influence of a seeded area on the other area, a buffer 'untouched' zone was designated between the two target areas.

The analysis of the Experiment I rainfall data made use of various statistical methods all of which indicated that there was a rainfall increase in the downwind target area: e.g., the root-double ratio method showed an increase of 15 per cent (standard error 6.5 per cent) for the entire area, whereas for the interior subsections (about 15 km inland, see Figure 10.1), the increase reached 22 per cent (6.7 per cent). Theoretical calculations show that under typical wind conditions, the area between 20 and 50 km downwind of the seeding route had the maximal exposure times to the optimal concentration of AgI particles (Gagin 1986).

In Experiment II (1969–75) the emphasis was on the Lake Tiberias catchment. The 'Central' target area was still seeded under the randomised predetermined procedure. Since this experiment took place after 1967 changes in governmental territorial jurisdiction, the central route (called the 'southern route' in Experiment II) could be extended southwards (broken line in Figure 10.1). In Experiment II the distance between the two routes was greater, therefore the 'crossover' technique of seeded–control relationship could not be applied. Only the northern target area rainfall data could be analysed, where the upwind northern coastal plain was designated as a non-seeded control area.

Experiment II statistical results are presented for eight subareas where the highest rainfall enhancement was 27 per cent (standard error 10 per cent), 40 km downward of the seeding line, while within the first 20 km nearest to the seeding line, the rainfall increased by 6–7 per cent (4–6 per cent) (Gagin and Neumann 1981).

Enhancement of rainfall due to the intensive urbanisation in the Coastal Plain of Israel has been studied employing three different methods. The most popular one is the comparison of long-term means ('normals'). Elbashan (1965) compared 1901–30 and 1931–60 rainfall normals. He mapped the percentage changes, which revealed that while in most of Israel there was a rainfall decrease, 15 km downwind of Tel Aviv there was an increase up to 28 per cent. These normals, as well as those for 1921–50 data for 18 stations in the Central Coastal Plain, were compared in a previous study (Goldreich and Manes 1979) which showed an increase of 5–17 per cent in three decades over Greater Tel Aviv and the downwind area.

The second approach for showing the urban influence on rainfall is to apply trend techniques. Goldreich and Manes (1979) used the Mann–Kendall rank statistics to show that in some urban and downwind weather stations there was a significant or nearly significant trend of rainfall increase.

Rainfall data

Annual sums of the daily rainfall amount for the seeded days were calculated for 57 quite well-distributed rainfall stations (with relatively fewer stations next to the coastal line) in the Central Coastal Plain (between the local latitude co-ordinates 114 and 194, see Figure 10.1). By 'seeded day' we mean that the 'central' target (or the 'southern' target in the second experiment) was seeded by planes. In contrast to the statistical analysis of the cloud-seeding experiment a 'seeded day' is considered only if actual seeding took place. On the other hand, a rainy day was considered as a 'seeded day' even if only a very short period of seeding was practically employed. Moreover, no differentiation was made between seeding routes; in a few cases in the second experiment inland routes were used rather than the usual offshore one. Since 1976 seeding is operational, and therefore the data in the present study are from 1961–75 only.

The 57 rainfall stations chosen were all the stations which had regular daily readings and not more than one year of observations were missing. The quotient reduction method was applied to restore missing data using nearby stations. The mean height of the stations is 52 m (the highest station is 125 m) and the mean distance from the sea is 9.5 km (the furthest station is 23 km). A subset of the original 57-station data set with 37 stations was defined (between latitude co-ordinates 122 and 174) to be representative of only the area downwind of the Central seeding route (Figure 10.1).

The multiple regression equations

The regression equations were computed (applying the stepwise procedure for the whole experimental period (Table 10.1) as well as for both experiments separately (Tables 10.2 and 10.3). For every sample and subsample (in size or period), an equation is presented where the dependent variables are the total rainfall amount, the total 'seeded days' and the total 'unseeded days' rainfall. The latter was calculated by subtracting the 'seeded' rainfall amount from the 'all' rainfall data. The independent variables were described above. Since there is a high correlation between height and distance variables (r is 0.74 and 0.76 for the 57- and 37-station samples, respectively), in some of the regression equations the height parameter substitutes the distance one. However, the height parameter is not pertinent to our discussion, therefore, when the height parameter entered the equation, the regression was computed again without the height. In most of these cases the distance parameter entered the equation with somewhat a lower correlation coefficient. In none of the equations do both of these related parameters appear together. The columns r and R present the simple correlation coefficients and the multiple correlation coefficient, respectively. The latitude (N) parameter always enters first into the regression, therefore the r column describes the relation between the rainfall parameters and the latitude, and the difference between r and R describes the additional contribution of the distance (D) (or height H) to the multiple regression. In cases where only one independent variable enters the regression equation significantly, the R column is left blank.

In the last column in the three tables the standard error (SE) of the distance (or height) regression coefficient is presented. This statistic was used as a significant test of the regression coefficient: the SE should be less than half of the regression coefficient magnitude to be considered significant at 95% for a two-tailed test. In other words, for such a significant level, if the SE is greater than half of the regression coefficient, this coefficient might appear in an opposite sign. The SE values are presented only slightly too large to meet the criteria for 95% confidence. Otherwise, in cases where SE was too high, the additional regression coefficient was rejected.

Table 10.2 As Table 10.1 for the first randomised seeding experiment

Period	No. of stations	Dependent variables	r	R	α	N	D	H	SE (D) or SE (H)
1961–7	57	All	0.88	0.89	81	2.9		0.39	
1961–7	57	Seeded	0.50		119	0.4			
1961–7	57	Not seeded	0.92	0.94	−40	2.4		0.4	
1961–7	57	Not seeded[1]	0.92	0.93	−42	2.5	1.5		
1961–7	37	All	0.90	0.92	−19	3.6		0.5	
1961–7	37	All[1]	0.90	0.91	−22	3.7	1.8		
1961–7	37	Seeded	0.74		55	0.9			
1961–7	37	Not seeded	0.91	0.93	−63	2.6		0.4	
1961–7	37	Not seeded[1]	0.91	0.93	−67	2.7	1.4		

Note: 1. See Table 10.1

Table 10.3 As Table 10.1 for the second randomised seeding experiment period

Period	No. of stations	Dependent variables	r	R	α	N	D	H	SE (D) or SE (H)
1969–75	57	All	0.78		303	1.9			
1969–75	57	Seeded	0.77		86	1.1			
1969–75	57	Not seeded	0.71	0.73	203	0.8		0.17	0.0097
1969–75	57	Not seeded[1]	0.71	0.73	199	0.8	0.72		0.46
1969–75	37	All	0.84		206	2.6			
1969–75	37	Seeded	0.88		−5.5	1.8			
1969–75	37	Not seeded	0.66		212	0.8			

Note: 1. See Table 10.1

Results and discussion

In the following discussion, the normals for 1951–80 are compared with means for the period 1931–60. The data for each of the seeding experiments are discussed thereafter.

Some support for the appearance of the distance variable in the equation of the 1951–80 sample can be detected from the rainfall map. In Figure 10.2 the isohyets of both rainfall normals are delineated. At a first glance both isohyet patterns look the same, especially if one compares the isohyets of 400 and 500 mm. However, while the isohyets of the former period tend to display a zonal direction, the isohyets of the latter one manifest a more meridianal orientation where the rainfall increases eastward (inland). This pattern change is more pronounced in the northern part by the 550 mm isohyet of the later period.

In a shorter period more stations with full data are available, so it was possible to analyse a larger sample (57 stations) for the experimental period (1961–75). The important result apparent from Table 10.1 is that only in 'not seeded' samples does the distance parameter enter the regression equation significantly with positive sign. We can therefore conclude that the intensive urbanisation of the last three decades is probably the primary cause of rainfall increase downwind of the coastline and built-up area. This conclusion conforms with previous studies (Elbashan 1965; Goldreich 1981; Goldreich and Manes 1979) in the relevant area

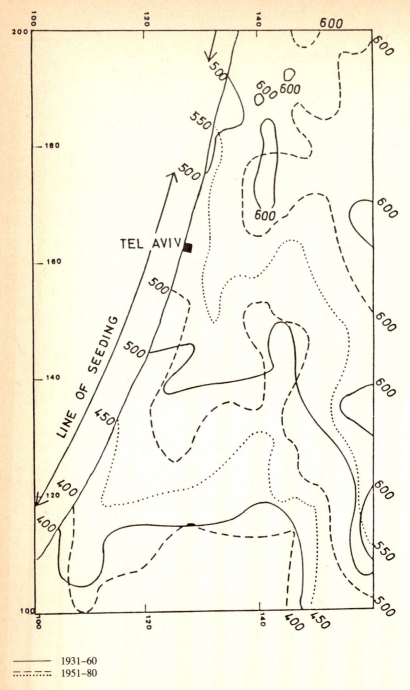

1931–60
1951–80

The grid is in local coordinates (km)

Figure 10.2 Isohyets (mm) of both rainfall normals for the central coastal plain and inland

in Israel as well as in other urban areas in the world.

This conclusion raises a question regarding the seeding effect. The same analysis might bring us to reason that there is no seeding effect in the Central Coastal Plain. Moreover, since the urban effect obviously exists during 'seeded' and 'not seeded' days alike, one might deduce that the seeding actually causes a negative effect.

However, both experiments showed by different methods that there is indeed a rainfall increase (although not a significant one) in the first 20 km downwind of the seeding line. Moreover, there is a high correlation (+0.87) between run-off data and rainfall amount for non-seeded days and a very low one (+0.20) for seeded days (in a small basin during Experiment II in the Northern target area) (Ben-Zvi 1985). The interpretation of this hydrological finding is that the seeding process leads to great spatial variance in rainfall, and the very localised showers affect the run-off in a spatially non-uniform pattern. This conclusion suits the difference in the R values for the 57 sample case but less so in the 37 one.

There is another hint supporting a positive influence of cloud-seeding. Consider that in Experiment I a 'seeded day' in the south is a 'non-seeded day' along the northern route and usually vice versa. Since the 57-station sample includes also a portion of the northern target area (or at least its influence), its r for the 'seeded' case (Table 10.2) is relatively low in comparison to the 'all' and 'not seeded' data sets, and also in comparison to the 37-station sample. This is in contrast to the 'not seeded' case where the smaller sample shows a lower R than the larger one. In the 'not seeded' 57 sample, the northern period was mostly seeded, a fact which probably increases the rainfall amount and consequently improves the linearity of the north–south rainfall distribution thus increasing its r and R values.

Conclusions

The differences between rainfall normals for two 30-year periods and the analysis for the whole experimental period lead to the conclusion that there is a trend of change in the isohyet map due to increases in urbanisation. This trend is manifest by an extraordinary spatial pattern where the rainfall increases with increasing distance from the sea in the Central Coastal Plain of Israel. In practical terms, the rainfall addition at 23 km (the furthest station) from the sea is between 5 and 15 per cent, depending on the spatial annual rainfall mean for the sample and the relevant regression coefficient value. The trend is not discerned during seeded days probably due to the spatial inhomogeneity of the seeding influence.

While this study gives vague support to the advertent effect that seeding has on rainfall near the seeding route, the inadvertent modification of rainfall due to urbanisation is more convincing.

Acknowledgements

The author would like to thank Miss A. Danziger for collecting and loading the data on the computer terminal. The data were supplied by the Israeli Meteorological Service. This study was partially sponsored by the Dr Irving & Cherna Moskovitz Chair in Israel Studies.

References

Ben-Zvi, A. 1985. 'A change in the daily rainfall-runoff model of a small basin due to cloud seeding', *Proc. IAHS Symp. on Scientific Basis for Water Resource Management*, Jerusalem, IAHS Publ. no. 153, 31–42.

Braham, R.R. Jr. 1981, 'Designing Cloud Seeding Experiments for Physical Understanding', *Bull. Amer. Met. Soc.* 62, 55–62.

Elbashan, D. 1965. 'Changes in the mean annual rainfall', *Meteorologia BeIsrael* 2, 42–43 (in Hebrew).

Gagin, A. 1986. 'Conceptual evaluation of "static" and "dynamic" seeding modes based on recent analyses of Israeli II and FACE II Experiment', *AMS Monog. on Glaciogenic Seeding for Precipitation Enhancement*, ch. 13.

Gagin, A., Neumann, J. 1974. 'Rain stimulation and cloud physics in Israel' in W.N. Hess (ed.) *Weather and Climate Modification*, New York, Wiley-Interscience, 454–94.

Gagin, A., Neumann, J. 1981. 'The second Israeli randomized cloud seeding experiment: evaluation of the results', *J. Clim. Appl. Met.* 20, 1303–11.

Goldreich, Y. 1981. 'The urban effect as an additional factor determining rainfall spatial distribution in Israel', *Isr. Meteorol. Res. Papers* 3, 193–202.

Goldreich, Y., Manes, A. 1979. 'Urban effect on precipitation patterns in the Greater Tel Aviv Area', *Arch. Met. Geoph. Biocl.* Ser. B, 27, 213–24.

Kerr, R.A. 1982. 'Cloud Seeding: One Success in 35 Years', *Science*, 217, 519–21.

Tukey, J.W., Brillinger, D.R., Jones, L.V. 1978. *The Role of Statistics in the Weather Resources Managements*. Vol. 2, Final Report of Weather Modification Advisory Board, Dept. of Commerce, Washington D.C.

Wolfson, N. 1975, 'Topographical Effects on Standard Normals of Rainfall over Israel', *Weather*, 30, 138–44.

Chapter 11

Trends in Maltese rainfall: causes and consequences

A.H. Perry University College of Swansea, United Kingdom

Introduction

It is fortunate that in the Mediterranean basin there is a number of meteorological stations where reliable precipitation records exist for long periods of years, and at a number of these sites studies have shown that large variations of annual precipitation totals have occurred. Colacino and Purini (1986) found that at Rome in the period 1782–1978 a number of dry and wet cycles could be recognised, while, at the eastern end of the basin, Plassard (1973) found that at Beirut a study of the record from 1876–1972 suggested that there may be cyclical trends in precipitation. To the author's knowledge, while these and other studies point to important long-term trends in precipitation, there have been no attempts to compare trends at different sites or to seek explanations for them in terms of known atmospheric changes that have occurred over Europe. The present chapter will firstly examine the precipitation record at Malta this century and then compare the observed variations with those noted elsewhere. Finally a number of possible causal mechanisms to explain the variations will be suggested.

The island of Malta has reliable rainfall recordings extending back to the middle of the nineteenth century. As a small island in the central Mediterranean without any large mountain massif (the highest gradient only just exceeds 200 m) Malta can be considered to exhibit a maritime variant of the Mediterranean climate and to be only affected by neighbouring land masses to a limited degree. Goosens (1985) has shown in a study of regionalisation of the Mediterranean area using annual precipitation data that Malta can be grouped with Sicily.

Unfortunately the Maltese rainfall record has to make use of data from different locations, but Bulmer and Stormarth (1960) produced a homogeneous record for the years 1854–1953. The Meteorological Office record at Luqa commenced in 1947 and, because of its reliability, it is this record which is used in the present paper for the last 40 years of data (1947–86). A comparison between the Luqa record and the mean of the gauges used before 1947 was carried out by Bulmer and Stormarth and shows a high degree of correlation in the observed trends. This is significant for the present paper which concentrates on analysing the trend of precipitation rather than actual amounts recorded. The analysis is restricted to the present century.

Rainfall trends in Malta, 1900–86

Ten-year running means were calculated for the 'homogeneous record' (1900–53) and similar running means based on the Luqa record were calculated for the period

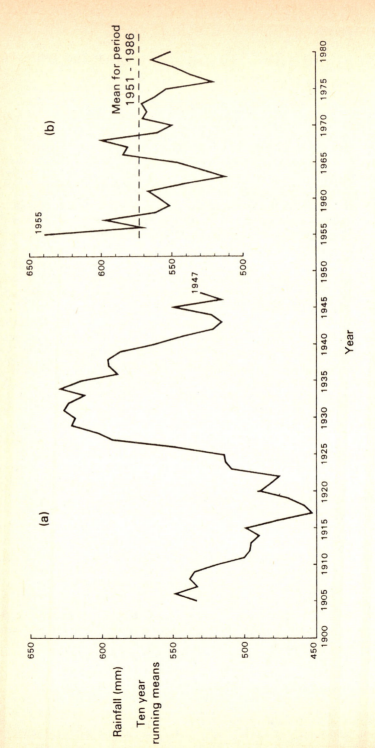

Figure 11.1 Ten-year running means for the (a) 'homogeneous record' (1900–53) and the (b) Luqa record (1951–86) on the right

of available data 1951–86. From Figure 11.1 it can be seen that the analysis suggests that there have been significant variations in the annual precipitation this century.

A dry period in the late 1910s was followed by an increase in rainfall that peaked in the 1930s. A further decline in the mid-1940s was followed by a very wet period in the early 1950s, since when there have been two dry periods in the early 1960s and the mid-1970s separated by a short wetter interval in the late 1960s.

As at all Mediterranean stations the bulk of the total annual rainfall falls in the winter months, with December on average the wettest month, with an average of 99.5 mm — 17.2 per cent of the annual total — at Luqa in the period 1951–86. The wet season lasts from September to April and outside these months only occasionally is rainfall measured, the most notable occasion being in August 1964 when 155 mm was recorded. Not surprisingly when ten-year running means of wet season rainfall were plotted they shared a high correlation with the annual running means. As measured by the coefficient of variation, variability of rainfall is least in December and increases progressively in both directions towards the dry season. Another feature of Maltese rainfall shared by other Mediterranean areas is the occasional extremely wet period at the beginning of the rainy season. The wettest month on record is October 1951 when 476 mm was recorded, a total which considerably exceeds that recorded in the driest individual year in the record, 1947, which had just 199 mm. Thus it can be concluded that the presence or absence of rain early in the wet season can be a major determinant in influencing individual annual totals. Although variability of rainfall in April is slightly greater than in September as measured by the coefficient of variation, the wettest April this century, 1955, had only 113 mm, so the spring months of March and April play a relatively minor role in influencing the annual total.

Rainfall trends at other Mediterranean stations

In order to answer the question of the extent to which the observed trends at Malta are typical of other long-period Mediterranean stations, a comparison will be made with records at Rome and in Corsica, both in the central Mediterranean, and at Beirut, in the eastern Mediterranean.

At Rome it was generally wet from the beginning of the century until the early 1940s, after which it was dry until the mid-1950s, since when, after a short wet period in the early 1960s, precipitation values have remained below those recorded in the first 30 years of the century. Thus there is little similarity with the Malta record overall, although at both places the post-1970 period has been rather dry. Orsini and Portrait (1986) have reported that in north-east Corsica the 1901–30 period was relatively wet, 1950–70 included many dry years, the 1970s were wet while since 1980 there has been a return to dryness. At Beirut there were wet periods in the first 20 years of the century and from the late 1930s to about 1960, while the 1920–40 period was very dry. The wettest period at this station has been since the 1960s.

The suggestions that there are wet and dry cycles at individual Mediterranean stations is widely found in the literature. At Rome, Colacino and Purini (1986) detect an oscillation with a period of about a century while at Beirut, Plassard (1973) suggests that the record exhibits 35- and 70-year oscillations. At Malta the precipitation minima in 1917, 1947 and 1977 are 30 years apart.

We can conclude that while marked dry and wet phases occur at many Mediterranean locations they rarely show a high degree of spatial coherence. This may be, of course, because it is in the nature of the Mediterranean climate that rain falls heavily on a relatively small number of days, and a few high daily totals can greatly influence the annual total. On the other hand, it is well known that Mediterranean precipitation is related to synoptic scale disturbances, the spatial scale of which is normally of the order of hundreds of kilometres.

Synoptic models, circulation changes and rainfall fluctuations

On the daily time-scale wet weather in the Mediterranean is associated with blocked circulation in middle latitudes, and especially when the tips of cold troughs become cut off to form cold pools. The relationship between wetness and blocked periods can also be observed at the monthly, and occasionally at the seasonal, time-scales. For example, November 1983, a month with persistent north Atlantic blocking, produced record rainfalls in southern Europe in an area extending from Barcelona in the west (which had 554 per cent of normal rainfall) through Malta (257 mm, compared with the average 98 mm) to Greece in the east. In 1976, when blocking patterns were extremely persistent, Malta had the second wettest year of the century, with some rainfall being measured in every month of the year. It might be expected that we could extrapolate from these relationships and therefore expect that phases of dominantly blocked circulation (post-1940) would be wet in the Mediterranean while phases of zonal circulation (pre-1940) would be dry. Among the explanations of why this is not so are first, that blocking is most frequent in the spring when, as has been noted, Mediterranean rainfall is less important than in autumn and winter; and second, depending on the criteria adopted to define blocking, it may be that the model of zonal and blocked epochs is too simple. For example Brezansky, Flohn and Hess (1951) found evidence of a 22–23-year oscillation in blocking over the period 1881–1950 with maxima in 1915 and 1937.

Studies of monthly and seasonal pressure patterns over the European sector of the northern hemisphere clearly show that in the 'blocked' post-1950 period, not all months of the year have been affected to the same extent. In particular October stands out as a month when zonal circulation has been particularly strong, a factor that contributed to an unusual prevalence of warm months over the British Isles. Thus situations like that in 1951, when a long period of low zonal index in late September and early October led to unprecedented rainfall in Malta, have occurred only rarely.

Consequences and conclusions

In Malta, as in many other island Mediterranean communities, the demands on limited water supply have increased in recent years, with more irrigated agriculture, and particularly as a result of the growth in tourist numbers. During a number of dry years in the 1970s Malta had to impose severe water restrictions in holiday resorts which led to a detrimental impression of the island by some visitors. By the early 1980s, as it became apparent that visitor numbers were falling, it was decided to go ahead with the building of two desalination plants, in order to get around the problem of capricious rainfall and uncertain water supplies.

Certainly, as can be seen from this chapter, it is extremely difficult to give any forecasts of the likely trends of rainfall in future years. Studies of rainfall this century do not show the clear relationship between circulation state and rainfall trend that might be inferred. This is because a few heavy daily falls frequently account for a large percentage of the total annual fall, and in the period September–November in particular even a short interval of low index circulation can lead to very heavy rainfall. This situation is in contrast to mid-latitudes where, as in the UK, for example, the influence of anomalous circulation can often be seen in the annual precipitation anomalies and even in 35-year period averages.

References

Brezansky, H., Flohn, H., Hess, P. 1951. 'Some remarks on the climatology of blocking action', *Tellus* 3, 191–4.

Bulmer, B.F., Stormarth, K. 1960. *The Rainfall in Malta*, Met. Office Scientific Paper no. 3, London, HMSO.

Colacino M., Purini R. 1986. 'A study of the precipitation in Rome from 1782 to 1978', *Theor. Appl. Climatol.* 37, 90–6.

Goosens, C. 1985. 'Principal component analysis of Mediterranean rainfall', *J. Climatol.* 5, 378–88.

Orsini, G., Portrait, J.F. 1986. 'Aléas climatiques, actions anthropiques et risques naturels dans une montagne Mediterranée: exemple de la Corse du nord-est', *Colloque climats et Risques Naturels*, Association Française de Géographie Physique, Paris.

Plassard, J. 1973. 'Studies of variability of the annual rainfall at Beirut 1876–1972', *Ann. Mem. of the Observ. de Ksara* 3(1).

Chapter 12

The recent variability of precipitation in north-western Africa

A. Douguedroit   University of Aix-Marseille II, France

The present drought has focused the interest of many scientists on the climate of Africa, but only a few of them direct their attention to the north-western area. Also they consider it as part of a larger area such as Africa (Nicholson, 1979; 1980) or the Mediterranean basin (Delannoy and Douguedroit 1983; 1984; Maheras 1985). One of the main reasons lies in the lack of a common long series of data. A first attempt at studying north-western Africa only was made recently (Douguedroit 1986).

The climate of north-western Africa is typically Mediterranean, with a dry summer lasting from three months in the north to five or more in the south. The amount of annual precipitation diminishes in the same direction from around 800–1000 mm to less than 100 mm, the southern area being quite dry. The more rainy stations situated in the mountains are not taken into account except Aïn-Draham (Tunisia) because their data are missing.

Most of the monthly data of the stations used were obtained from the National Center of Atmospheric Research (Ashville, Alabama, USA) and the monthly or daily publications of the Meteorological Service of Morocco and Tunisia. A few gaps were filled by a multiple regression analysis.

The model

Principal Component Analysis (PCA) has been used by a few scientists to analyse rainfall patterns, the first of these being Steiner (1965). He applied it to a set of data for 16 variables including annual, January and July precipitation, for 67 stations in the United States.

The two-way variance model is an extended version of PCA (Antoniadis *et al.* 1986). It is expressed as follows:

$$p_{ij} = p_j + \sum_{k=1}^{q} \alpha_{kj} \beta_{ki} + \epsilon_{ij}$$

where

p_{ij} is the rainfall amount at station j for the time i
p_j is the average rainfall at the station j
q is the number of structured components retained to explain the variance
α_{kj} is the space coefficient at station j for component k; it measures the influence of component k on the station
σ_{ki} is the time coefficient of component k at the time i
ϵ_{ij} is the residual term which represents the part of the anomaly not represented by the model.

As the q time series are orthogonal, the first one usually represents, as in PCA, a common history of the whole area and the others represent apparent spatial contrasts. To free the orthogonality constraint on space coefficients, after selecting the q functions, an orthogonal rotation is applied on the time coefficients according to an 'OBLICIMIN' criterion (Feyt 1987). The total percentage of variance remains the same but it is combined in a different way; the most important part of the represented variance at each station tends to cluster on only one component. Each component includes stations with common time variability, situated in a homogeneous area from the point of view of the inter-annual variability of their precipitation. It consists in the variations of the influence of the component on the station it influences most. It is considered to represent the time variability of the whole region concerned.

Two-way variance model with rotation of north-western Africa rainfall

As the model necessitates the use of uninterrupted series, we could only select 62 stations with different lengths of records (Table 12.1). We processed several

Table 12.1 List of stations used for variance analysis

Agadir (M) 1922–71	Laghouat (A) 1922–60
Aïn-Draham (T) 1922–80	Le Kef (T) 1951–80
Aïn-el-Hadjar (A) 1925–50	Marrakech (M) 1922–71
Aïn-Sefra (A) 1922–60	Mateur (T) 1951–80
Algiers (A) 1922–71	Medenine (T) 1922–80
Azrou (M) 1922–60	Meknes (M) 1922–81
Barrage-el-Khebi (T) 1951–80	Meskiana (A) 1925–81
Bechar (A) 1922–71	Misurata (L) 1931–71
Beja (T) 1951–80	Mohammedia (M) 1922–60
Berkane (M) 1922–60	Moktar (T) 1951–80
Biskra (A) 1922–71	Monastir (T) 1931–80
Casablanca (M) 1922–71	Nalut (L) 1922–60
Chelghoum (A) 1925–50	Oran (A) 1922–73
Constantine (A) 1922–71	Ouedzem (M) 1922–60
Djerba (T) 1931–80	Oujda (M) 1922–71
El Djem (T) 1951–80	Rabat (M) 1922–60
El Eulma (A) 1925–50	Safi (M) 1931–71
El Feidja (T) 1951–80	Settat (M) 1922–60
El Golea (A) 1922–71	Sfax (T) 1922–80
El Jadida (M) 1922–71	Sidibennour (M) 1922–71
El Oued (A) 1922–71	Si Kassem (M) 1922–71
Essaouira (M) 1922–60	Tabarka (L) 1931–80
Gabes (T) 1951–81	Tangiers (M) 1922–71
Gafsa (T) 1922–80	Taza (M) 1922–71
Ghardaïa (A) 1922–71	Tebessa (A) 1922–60
Guercif (M) 1922–60	Teboursouk (T) 1951–80
Jendouba (T) 1951–81	Tlemcen (A) 1922–71
Kairouan (T) 1922–80	Tozeur (T) 1931–80
Kasserine (T) 1951–80	Tripoli (L) 1922–80
Kenitra (M) 1922–60	Tunis (T) 1922–80
Khouribga (M) 1925–60	Youssoufia (M) 1931–60

Key: A = Algeria
    L = Lybia
    M = Morocco
    T = Tunisia

analyses according to the length of the series and the geographical regions but including a few common stations. One group spread all over the country extended in time from 1931 to 1972. Second and third groups from 1922 to 1960 and from 1925 to 1950 allowed us to take into account a greater number of dry southern or eastern stations in Algeria. A fourth one included only 21 stations in Tunisia from 1951 to 1980 to test the results obtained by the former analysis including only ten of them.

For each group of stations, the two-way variance model with rotation was applied to the monthly data of every rainy month (from January to May and September to December), of four seasons (autumn, winter, spring and the rainy season from October to April), to the totals of the 12 months taken one after the other during the aforementioned years and to the yearly totals.

Spatial pattern

We collected the results of the three types of time analysis previously pointed out for each month or season and for the year to sum up the spatial pattern of the variability of the precipitation in the whole area. We used the results of the analysis of the 1922–60, 1925–50 and 1951–80 series to clarify the limits of the regions coming from the 1931–72 series.

The network of stations was not evenly distributed. In some areas, data were entirely missing. In the others, they displayed an understandable regional pattern, with rather well determined cores and more uncertain boundaries (Figure 12.1).

The general layout of the spatial pattern does not change during the rainy season. It displays meridianal and latitudinal limits, many of which are in keeping with geographical features such as high mountains.

The Middle-Atlas and High-Atlas isolate western Morocco, the regions of which are always well defined. They are formed whatever the number of components is in the analysis. They are made up of stations with very similar variability of precipitation and high percentage of variance represented by common components after rotation. Western Morocco can be divided into four regions, from north to south: a small far-northern region around Tangiers, a northern one, a central and a southern region (regions 1–4 in Figure 12.1). During some months, differences appear between the coastal and inner areas (in the centre in January and in the north in November).

East of the Middle-Atlas, the main divisions are latitudinal according to the relief. Three strips acquire an identity of their own, from the north to the south: the coastal, the high plains and the southern areas. The sharpness of their limits depends on the difference in level. They are plain in Algeria but they fluctuate according to the time of year in the steppe area of Tunisia.

The coastal strip presents an overall unity from Oujda to Tunis. With a few components, all the stations spread along or near the coast cluster on only one component which mostly represents Algiers (80 per cent variance) and secondarily the other stations (50 per cent or less variance). When the number of components increases, several groups of stations depart from the unity and group together on specific components with a high percentage of variance. So we can consider that the coastal strip is divided into five areas: north-western Algeria and north-eastern Morocco with an uncertain western limit because of the lack of stations along the Moroccan coast, north-central Algeria with Algiers, north-eastern Algeria with

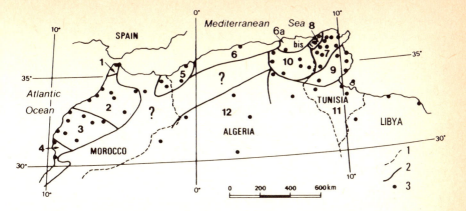

Figure 12.1   Spatial pattern of precipitation in north-western Africa

1 = Frontier
2 = Boundary of homogeneous area
3 = Station

Constantine, the mountainous region of the frontier and the north-eastern Tunisia area which extends towards the south as far as the Tunisian Dorsale (regions 5, 6, 6a, 8 and 7, respectively in Figure 12.1). Some uncertainty concerns eastern Algeria where data are missing.

Within the country, a medial area stretches between the Middle-Atlas and the Tunisian coast. Only the eastern part is known; it is divided into two zones, including the steppe area in Tunisia and separated by a rough limit (regions 9 and 10 in Figure 12.1).

The southern area begins at the foot of the Moroccan-Algerian Atlas mountains down to the Sahara. It is dry, with less than 200 mm annual rainfall and a Mediterranean regime of precipitation. It can be divided into two parts, the east and the centre, the western limit of which is unknown.

Here this schematic general pattern is altered during some months. At the beginning and the end of the rainy season, the Tunisian area stretches as far as Libya. In contrast, during the heart of the same season the differences between central and southern Tunisia, which are fairly flat regions, become blurred. The rainfall fluctuations are more progressive than sudden. So limits induced for a climatic regionalisation can vary from one month to another or between seasons.

Time variability

The inter-annual variability of each component represents the time evolution of a whole area and a part of represented variance at each station. The station with the best-represented variance and the area in which it is included have basically a similar inter-annual variability. According to the hypothesis of steady climatic regions at the time-scale of a century, we can consider those stations with the best-represented variance and their time evolution as representative of a region during the years before and after those included in a variance analysis.

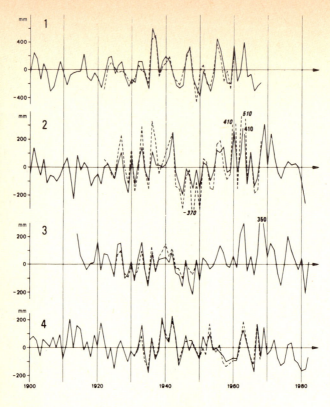

------ The components

———— The station with the most important variance represented

Figure 12.2   Inter-annual variability of annual precipitation

 1 = Tangiers: mean annual rainfall, 870 mm; variance represented, 72%
 2 = Casablanca: 420 mm and 85%
 3 = El Jadida: 340 mm and 73%
 4 = Essaouira: 280 mm and 84%
 5 = Tlemcen: 660 mm and 80%
 6 = Algiers: 455 mm and 81%
 7 = Tunis: 462 mm
 8 = Aïn-Draham: 1530 mm (scale on the right) and 75%
 9 = Kairouan: 280 mm and 75%
10 = El Eulman: 370 mm and 73%
11 = Medenine: 155 mm and 63%
12 = Aïn-Sefra: 168mm and 65%

We only study here the case of the annual precipitation. Some common features appear in the time variations of western Morocco, except in Tangiers (Figure 12.2) — rainy years about 1940 and a trend of diminishing precipitation since 1970 which should be studied more accurately and compared with the variations in the southern Sahara (we are investigating such a comparison).

The similarity between the central and the northern areas is more important — two trends of increasing and diminishing precipitation from 1944 up to the present,

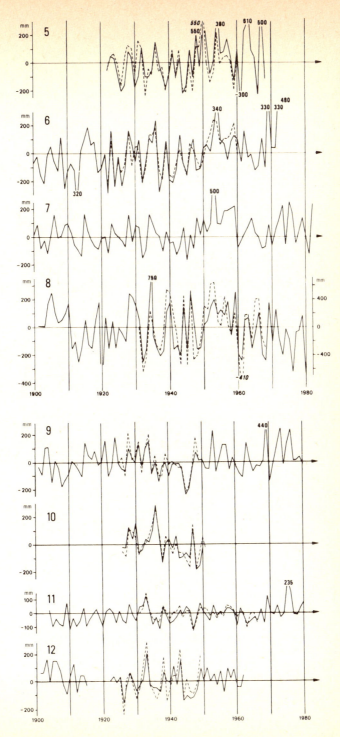

separated by four dry years (1964–1967). Their components are significantly correlated at the 95% level during their common period; but when all the stations of both areas are included in the same component, the central stations have only a rather weakly represented variance (less than 50 per cent), whilst before 1925 the diagrams of El Jadida and Casablanca differ. The rainfall evolution of the southern area displays dry periods in the 1920s and at the end of the 1950s that are unknown in the north.

The variations of annual rainfall appear quite different east of the Middle-Atlas. They present some similarities in all the areas of the coastal zone, except in mountainous western Tunisia. In these four areas they can be gathered in a single component but only the variance of Algiers is well represented. The represented variance of the stations of the other areas is less than 50 per cent. On Figure 12.2 diagrams 5, 6 and 7 display a kind of common increasing trend between 1920 (1910) and 1955–1960, but with differences in details and since 1960. In the mountainous boundary area, the mean departures become very important at the stations situated in altitude such as Aïn-Draham. Everywhere along the coast, it was rainy in the 1950s.

A few dry periods appear in the medial zone: at the end of the 1940s (central Tunisia and eastern Algeria) and at the beginning of the century (in Kairouan). In the southern zone, they appear during other periods: about 1910 at Medenine, at the beginning of the 1930s and the end of the 1940s in southern Tunisia and Algeria. A kind of increasing precipitation trend can be pointed out at Medenine from 1910 to 1930 (Figure 12.2).

Conclusion

This study of the annual precipitation of north-western Africa gives a first outline of variations which needs some improvement. The characteristic features of variations of precipitation pointed out during the twentieth century will be investigated more closely in the future.

References

Antoniadis, A., Charre, J., Degerine, S., Gregoire, G., Lebreton, A., Martin, S. 1986. 'Modèles d'analyse de la variance à deux facteurs pour l'études des précipitations sur un réseau de station', Research Report no. 584, IMAG, Grenoble.

Delannoy, H., Douguedroit, A. 1983. 'A propos des précipitations dans le sud-ouest européen et le Maghreb occidental', *La Météorologie* 34, 169–81.

Delannoy, H., Douguedroit, A. 1984. 'Les variations des précipitations printanières dans le sud-ouest européen at le Maghreb occidental (1916–1965)', *Rev. Géog. de l'Est* 1, 47–63.

Douguedroit, A. 1986. 'Le renouvellement méthodologique de la définition des climats régionaux', *Freiburger Geographische Hefte* 26, 268–79.

Feyt, G. 1987. 'Travaux méthodologiques pour l'analyse de la structure spatio-temporelle d'un phénomène climatique: application à l'étude des variations pluviométriques en Europe occidentale durant le siècle écoulé', Doctoral thesis, Grenoble.

Maheras, P. 1985. 'A factorial analysis of Mediterranean precipitation', *Arch. Met. Geoph. Biocl.* ser. B, 36, 1–14.

Nicholson, S.E. 1979. 'Revised rainfall series for the West African subtropics', *Mon. Wea.*

Rev. 107, 620–3.

Nicholson, S.E. 1980. 'The nature of rainfall fluctuations in subtropical West Africa', *Mon. Wea. Rev.* 108, 473–87.

Steiner, D. 1965. 'A multivariate statistical approach to climatic regionalization and classification', *Tijdischrifte van het koninklk nederlansch aardrijskundig genootschop* 82, 4, 329–47.

Part III
Tropical and southern Africa

Chapter 13

Modelling the influence of sea-surface temperatures on tropical rainfall

J.A. Owen and C.K. Folland Meteorological Office, United Kingdom

Introduction

The Sahel region of northern Africa (see Figure 13.1) has suffered from persistent drought in recent times (Dennett *et al.* 1985; Nicholson 1985). A time series of Sahel rainfall for this century (see Figure 15.1, this volume, p. 167) shows that rainfall has been below the long-term (1901–80) mean in every year since 1968. By contrast, the 1950s had persistently above average rainfall.

Many local factors affect the rainfall: albedo feedback (Charney 1975), soil moisture feedback (Walker and Rowntree 1977) and anthropogenic effects (Otterman 1977). This chapter, while not ignoring these effects, is mainly concerned with changes in the general circulation which are associated with reduced Sahel rainfall. It was originally thought that changes in the position of the Intertropical Convergence Zone (ITCZ) were responsible (Winstanley 1973; Lamb 1978). The idea has been disputed (Miles and Folland 1974), and it was suggested instead (Nicholson 1981) that a weakened intensity of the rainy season, independent of the ITCZ position, is the dominant cause of drought in the Sahel.

Newell and Kidson (1984) showed that, in drier Sahel years, the low-level southwest winds that bring moisture into west Africa were weaker and shallower in vertical extent than in wetter years, although they penetrated the usual distance inland. The same paper contains evidence of slightly reduced upper tropospheric easterly flow over the Sahel in dry years. Combined with weaker low-level westerly flow, this implies some reduction in vertical wind shear which can adversely influence the formation of easterly waves (Hastenrath 1985) that are responsible for much of the rainfall in the Sahel (Riehl 1979).

Sea-surface temperatures (SSTs)

Many general circulation model (GCM) experiments have demonstrated the effects of tropical SST anomalies (Palmer and Mansfield 1984; Shukla and Wallace 1983) and it has been shown that the inclusion of observed as opposed to climatological SSTs has a beneficial effect on the skill of some extended-range dynamical forecasts (Mansfield 1986; Owen and Palmer 1987).

A number of authors (e.g. Lamb 1978; Lough 1986; Hastenrath 1984) have suggested connections between Sahel rainfall and Atlantic SSTs. Folland, Palmer and Parker (1986) showed for the first time, however, that persistently wet and dry periods in the Sahel are strongly related to SST anomalies on a *near-global* scale. They found a correlation coefficient of -0.62 between the July to September SST difference between the southern and northern hemisphere (all of the Indian Ocean

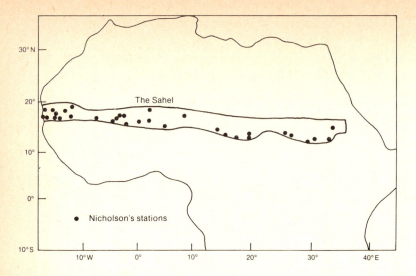

Figure 13.1   The Sahel region as defined by Nicholson (1985)

being included with the southern hemisphere) and Sahel rainfall for 1901–84. They carried out GCM experiments which demonstrated strong effects of a composite SST anomaly (the difference in SST between five dry and five wet Sahel years) on Sahel rainfall. Palmer (1986) also found fairly strong effects from SST anomalies in the Atlantic, Pacific and Indian Oceans, but the largest effect was from the composite global SST anomaly.

The aim of the present work is to model the effects of SSTs measured in individual wet (1950) and dry (1984) years. The years 1950 and 1984 were, respectively, the wettest and driest years in the Sahel this century. The latter was interesting in that the ITCZ over the Atlantic was unusually far south in that year (Philander 1986). In the model results we should, therefore, be looking for evidence of both a southward shift in the north African rainfall belt and changes in the large-scale tropical circulation.

We were also interested in assessing the importance of soil moisture feedback. We were able to do this since soil moisture content is a prognostic variable in our GCM, and we were able to include or exclude its variation.

Model and experiments

All the experiments described here were carried out with the Meteorological Office ll-layer GCM (Slingo 1985) which has a resolution of 2.5° latitude × 3.75° longitude. Two versions of the model have been used to obtain the results in this chapter. Some experiments used a model much like that described by Cunnington and Rowntree (1986) (model '3'), while others used model '4' which allows prediction of changing cloud amounts from model variables, incorporates a parametrisation of orographic gravity wave drag (Palmer, Shutts and Swinbank 1986) and uses an improved calculation of soil temperature, along with increased surface albedo in the Sahara.

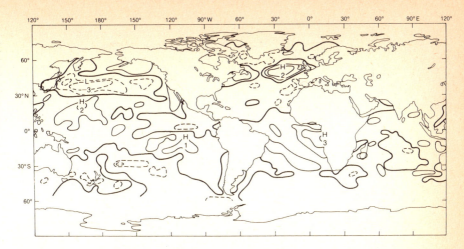

Figure 13.2   Difference in SST between 1984 and 1950 for July to September

Note: Contour interval is 1 °C with negative contours dashed.

Each experiment consisted of a 210-day integration in annual cycle mode, starting from the atmospheric conditions of 26 March 1984. The model uses 'months' of 30 days each, so the seven 'months' of each integration correspond approximately to April, May, . . ., October. This approximation will be used throughout the paper.

The SSTs used were obtained as monthly means from the Meteorological Office Historical Sea Surface Temperature data set and interpolated to five-day means. The SSTs were updated every five days during the course of each integration. Half of the integrations used 1950 SSTs, and the other used 1984 SSTs. The difference in SST between 1984 and 1950 for July to September is shown in Figure 13.2.

Soil moisture feedback in the model takes place through the interaction of the soil moisture content with precipitation, evaporation, run-off and snowmelt. This was allowed in experiments 7028, 7029, 7033 and 7034, but prevented in others. Table 13.1 lists the experiments.

Results of experiments including soil moisture feedback

Rainfall

Figure 13.3 shows the rainfall over north Africa for August from experiment 7028. The observed rainfall for August 1950 obtained from CLIMAT reports is shown in Figure 13.4. Apart from the details of the precipitation pattern in the coastal regions near the Gulf of Guinea and excessive rainfall in some parts of the Sahara, the simulation is remarkably good both as regards the pattern and amount of rainfall. The same is true of the July and September simulations though the July rainfall is generally underestimated and the September rainfall overestimated. Figure 13.5 shows the corresponding rainfall from experiment 7029, and the observed rainfall

Table 13.1 List of experiments

Experiment number	SSTs used	Version of model used	Interactive soil moisture	Soil moisture climatology
7028	1950	3	Yes	–
7029	1984	3	Yes	–
7030	1950	3	No	3
7031	1984	3	No	3
7033	1950	4	Yes	–
7034	1984	4	Yes	–
7035	1950	4	No	3
7036	1984	4	No	3
7037	1950	4	No	4
7038	1984	4	No	4
7040	1950	3	No	4
7041	1984	3	No	4

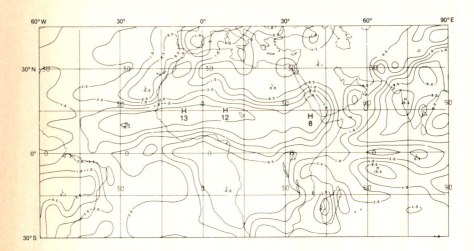

Figure 13.3   Modelled August rainfall for experiment 7028

Note: Contours are at 0.5, 1, 2, 4, 8 and 16 mm^{-1}.

for August 1984 is shown in Figure 13.6. Again there is excess modelled rainfall over parts of the Sahara, but generally the amounts are realistic.

Of particular interest is the difference in modelled rainfall between experiments 7029 and 7028, shown for July to September in Figure 13.7, with the observed differences between 1984 and 1950 in Figure 13.8. The differences in modelled rainfall are surprisingly realistic: the reduced rainfall over the Sahel and the increase farther south are shown. These results are consistent with the observation (Philander 1986) that the ITCZ was farther south than usual over the Atlantic in 1984.

The corresponding rainfall differences obtained using model 4 (7034 minus 7033) are shown in Figure 13.9. Similarities between this diagram and Figure 13.7 are evident, especially over the oceans, where there is generally reduced rainfall at around 15°N, and increased rainfall to the south of this latitude. Over Africa,

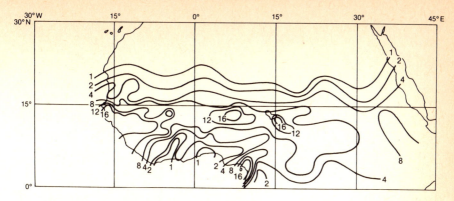

Figure 13.4   Observed rainfall for August 1950

Note: Contours are at 1, 2, 4, 8, 12 and 16 mm day^{-1}.

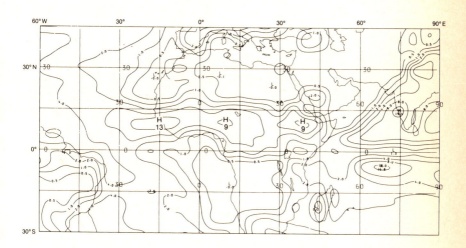

Figure 13.5   Modelled August rainfall for experiment 7029

Note: Contours as in Figure 13.1.

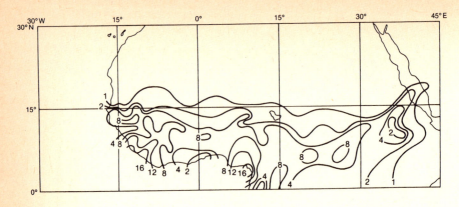

Figure 13.6   Observed rainfall for August 1984

Note: Contours as in Figure 13.4.

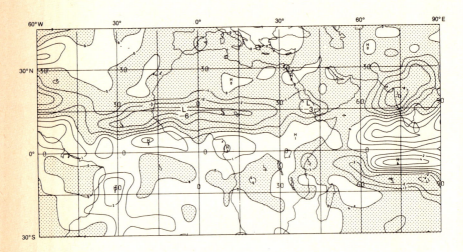

Figure 13.7   Modelled rainfall: difference between experiments 7029 and 7028, for July to September

Note: Contours are at 0, ±1, 2, 4, 8 and 16 mm day^{-1}, with negative areas stippled.

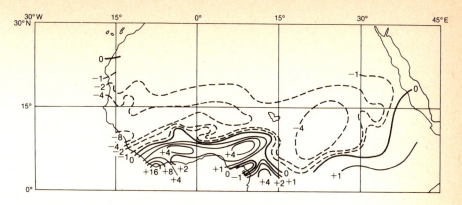

Figure 13.8 Observed rainfall: difference between 1984 and 1950, for July to September

Note: Contours as in Figure 13.7, with negative contours dashed.

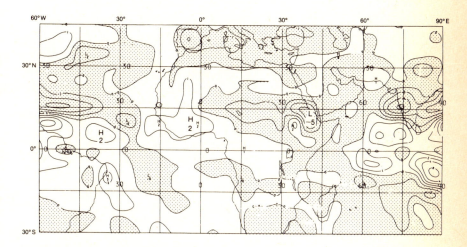

Figure 13.9 Modelled rainfall: difference between experiments 7034 and 7033, for July to September

Note: Contours as in Figure 13.7.

reduced rainfall over the eastern Sahel is clearly seen, but is not present over the western Sahel. One reason for this is that model 4 has a poorer rainfall climatology over the western Sahel than model 3, with too little rain. The lack of a substantial difference between 7034 and 7033 is a consequence of this failing.

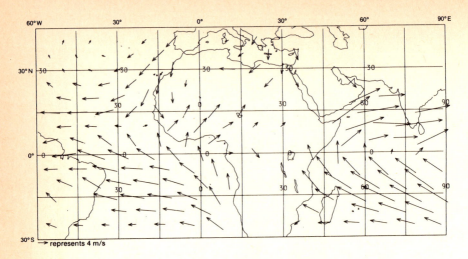

Figure 13.10   950 mb wind climatology for Model 3, for July to September

Winds

Figure 13.10 shows the 950 mb wind climatology of model 3 for July to September. The cross-equatorial flow over the Atlantic and south-westerly monsoon flow into western Africa are clearly seen. This flow is the primary source of moisture for the western Sahel, so that any reduction of this flow will have a major impact on Sahel rainfall.

Figure 13.11 shows the difference in 950 mb wind between experiments 7029 and 7028 for July to September. Over western Africa the 1984 SSTs have weakened the monsoon flow compared with 1950. The corresponding difference for the 250 mb wind is in a westerly sense over most of the tropics and indicates a reduction in the normal upper-level easterly flow. These results are consistent with those of Newell and Kidson (1984), although their analysis did not include the extreme years 1950 and 1984.

Experiments using model 4 gave similar results. In order to investigate the statistical significance of these wind changes, the months of July to October for each of the experiments 7028, 7029, 7033 and 7034 were split into three, supposedly independent, 40-day periods. Cross-section differences in zonal wind averaged over longitudes 10 °W to 30 °E between the mean of the six periods which used 1984 SSTs and the six periods which used 1950 SSTs are shown in Figure 13.12. Regions where the value of the Student's t-statistic is greater than 2.2 in magnitude are dotted. On the assumption that each mean comprises six independent values, such values of t indicate that the difference is significant at the 95% level of confidence.

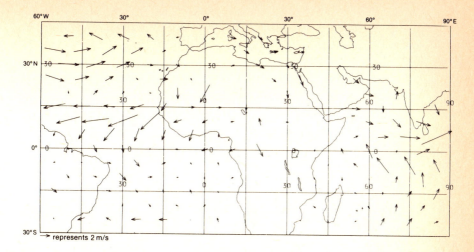

Figure 13.11 950 mb vector wind: difference between experiments 7029 and 7028

Note: change of wind-scale from Figure 13.10.

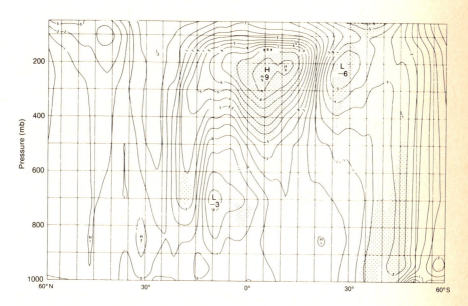

Figure 13.12 Cross-section (average over longitudes 10 °W to 30 °E) of the zonal wind difference between the July to October mean of experiments 7029 and 7034 and the July to October mean of experiments 7028 and 7033

Note: Contour interval is 1 ms^{-1}. Areas where this difference is significant at the 95% confidence level are stippled.

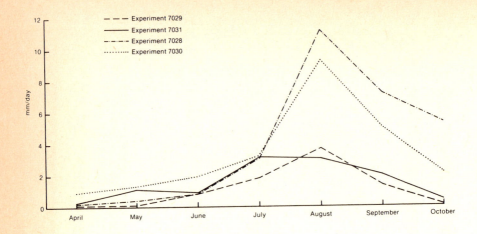

Figure 13.13   West Sahel rainfall time series for experiments 7028 to 7031

Soil moisture feedback

Soil moisture feedback has been mentioned as a potentially important effect. The experiments described in the previous section were rerun with the soil moisture everywhere prescribed to follow the climatology of model 3 or model 4. This gave eight experiments in which soil moisture feedback could not take place.

Figure 13.13 shows the rainfall time series for the western Sahel for four experiments (7028–7031) which used model 3. The 1950 SST experiments clearly produced more rain than the 1984 SST ones, even when soil moisture feedback is prevented from taking place. The results indicate, however, that this feedback mechanism amplifies the rainfall differences, especially towards the end of the integrations when the soil moisture differences are large (Figure 13.14 shows the west Sahel time series of soil moisture content).

The effects of soil moisture feedback on wind changes are slight. Figure 13.15 shows the zonal wind difference averaged between 10 °W and 30 °E between experiments 7031 and 7030 (climatological soil moisture). This is very similar to Figure 13.12 which was based on experiments with interactive soil moisture.

The effects on temperature are shown in Figures 13.16 and 13.17 for a more limited latitude band. The large difference in temperature over the Sahel is present only in the experiments with interactive soil moisture. Clearly soil moisture feedback strongly affects the local response to SST changes, but little affects the large-scale changes in general circulation.

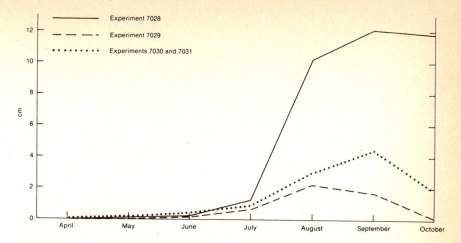

Figure 13.14 West Sahel soil moisture for experiments 7028 to 7031

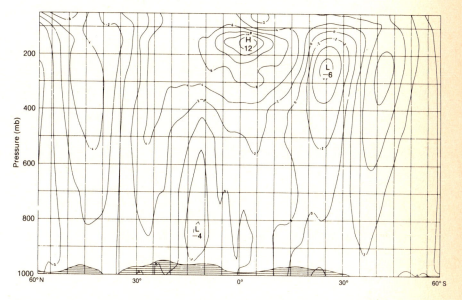

Figure 13.15 Cross-section (average over longitudes 10 °W to 30 °E) of the zonal wind difference between the July to September means of experiments 7031 to 7030

Note: Contour interval is 2 m s^{-1}.

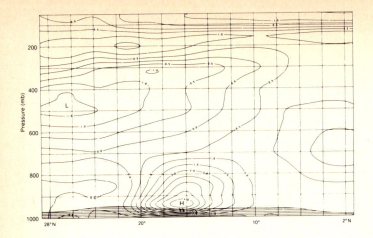

Figure 13.16   As Figure 13.15 but for temperature and experiments 7029 and 7028

Note: Contour interval is 0.5 °C.

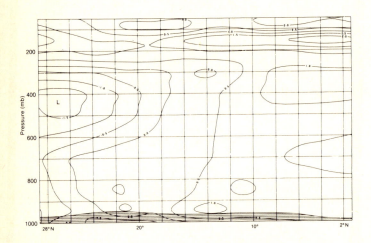

Figure 13.17   As Figure 13.16 but for experiments 7031 and 7030

Conclusions

The Meteorological Office 11-layer model has shown a clear ability to simulate not only the observed rainfall climatology of the Sahel but, more importantly, observed changes in Sahel rainfall in response to different sea-surface temperatures. Two versions of the model gave statistically significant changes in wind in response to the imposed SSTs, and these changes are consistent with observed wind changes between wet and dry periods in the Sahel. Soil moisture feedback in the model affects the local differences between the 1950 SST and 1984 SST integrations, but the SSTs seem mainly responsible for initiating these differences.

References

Charney, J.G. 1975. 'Dynamics of deserts and drought in the Sahel', *Q. J. R. Met. Soc.* 101, 193–202.

Cunnington, W.M., Rowntree, P.R. 1986. 'Simulations of the Saharan atmosphere — dependence on moisture and albedo', *Q. J. R. Met. Soc.* 112, 971–99.

Dennett, M.D., Elston, J., Rodgers, J.A. 1985. 'A reappraisal of rainfall trends in the Sahel', *J. Climatol.* 5, 353–61.

Folland, C.K., Palmer, T.N., Parker, D.E. 1986. 'Sahel rainfall and worldwide sea temperatures, 1901–85' *Nature* 320, 602–7.

Hastenrath, S. 1984. 'Interannual variability and annual cycle: mechanisms of circulation and climate in the topical Atlantic sector', *Mon. Wea. Rev.* 112, 1097–1107.

Hastenrath. S. 1985. *Climate and Circulation of the Tropics*. Dordrecht, D. Reidel.

Lamb, P.J. 1978. 'Large scale tropical Atlantic surface circulation patterns associated with sub-Saharan weather anomalies', *Tellus* 30, 240–51.

Lough, J.M. 1986. 'Tropical Atlantic sea surface temperatures and rainfall variations in Subsaharan Africa', *Mon Wea. Rev.* 114, 561–70.

Mansfield, D.A. 1986. 'The skill of dynamical long-range forecasts, including the effect of sea surface temperature anomalies', *Q. J. R. Met. Soc.* 112, 1145–76.

Miles, M.K., Folland, C.K. 1974. 'Changes in the latitude of the climatic zones of the northern hemisphere', *Nature* 252, 616.

Newell, R.E., Kidson, J.W. 1984. 'African mean wind changes between Sahelian wet and dry periods', *J. Climatol.* 4, 27–33.

Nicholson, S.E. 1981. 'Rainfall and atmospheric circulation during drought periods and wetter years in west Africa', *Mon. Wea. Rev.* 109, 2191–209.

Nicholson, S.E. 1985. 'Sub-Saharan rainfall 1981–84', *J. Clim. Appl. Met.* 24, 1388–91.

Otterman, J. 1977. 'Anthropogenic impact on the albedo of the earth', *Climatic Change* 1, 137–55.

Owen, J.A., Palmer, T.N. 1987. 'The impact of El Niño on an ensemble of extended-range forecasts', *Mon. Wea. Rev.* 115, 2103–17.

Palmer, T.N. 1986. 'Influence of the Atlantic, Pacific and Indian Oceans on Sahel rainfall', *Nature* 322, 251–3.

Palmer, T.N., Mansfield, D.A. 1984. 'The response of two atmospheric general circulation models to sea surface temperature anomalies in the tropical east and west Pacific', *Nature* 310, 483–5.

Palmer, T.N., Shutts, G., Swinbank, R. 1986. 'Alleviation of a systematic westerly bias in general circulation and numerical weather prediction models through an orographic gravity wave drag parametrization, *Q. J. R. Met. Soc.* 112, 1001–40.

Philander, S.G.H. 1986. 'Unusual conditions in the tropical Atlantic Ocean in 1984', *Nature* 322, 236–8.

Riehl, H. 1979. *Climate and Weather in the Tropics*, New York, Academic Press.

Shukla, J., Wallace, J.M. 1983. 'Numerical simulation of the atmospheric response to equatorial Pacific sea surface temperature anomalies', *J. Atmos. Sci.* 40, 1613–30.

Slingo, A. 1985. Handbook of the Meteorological Office 11-layer atmospheric general circulation model, *Dynamical Climatology Technical Notes no. 29*, Bracknell, Meteorological Office.

Walker, J., Rowntree, P.R. 1977. 'The effect of soil moisture on circulation and rainfall in a tropical model', *Q. J. R. Met. Soc.* 103, 29–46.

Winstanley, D. 1973. 'Recent rainfall trends in Africa, the Middle East and India', *Nature* 243, 464–5.

Chapter 14

Sea-surface temperature – Sahel drought teleconnections in GCM simulations

L.M. Druyan Bar-Ilan University, Israel and NASA, New York, USA

Introduction

Characteristic patterns of sea-surface temperature (SST) anomalies have been identified with droughts over north Africa, in particular in the Sahel. Lamb (1978) and Lough (1986) have described SST anomalies in the Atlantic Ocean that have preceded or have been concurrent with dry summers in the Sahel. Folland *et al.* (1986) described global SST anomaly patterns associated with these droughts and offered a general circulation model (GCM) experiment that simulates modest reductions in rainfall over north Africa by prescribing SST anomalies. In a series of 37 GCM-generated summer climates, Druyan (1987) found that computed SSTs in the South Atlantic preceding very dry summers in the Sahel were different from SSTs before rainy summers, the differences being similar to the anomalies cited by the observational studies. The same study also described synoptic differences between atmospheric conditions simulated by the GCM preceding 'model drought' over the Sahel and those preceding computations of copious Sahel rainfall. The results suggested that warm SST anomalies in the south-east Atlantic may cause reductions of precipitation over north Africa by lowering atmospheric pressures, thereby weakening the moisture-laden southerly flow along the south-west coast of Africa.

The present research continues the work with the GISS GCM described by Druyan (1987). It attempts to clarify the sensitivity of the model computations of summertime rainfall over the Sahel to the intensity, geographic configuration, and timing of the SST anomalies.

The GCM is model II described by Hansen *et al.* (1983), except that it uses the interactive ocean formulation described in Druyan (1987). The model computations are made on a coarse grid with horizontal resolution 8° latitude × 10° longitude with nine vertical layers. Physical processes that are modelled include solar and terrestrial radiation transfer; vertical transport of moisture, sensible heat and momentum by convection; clouds from supersaturation and convection; heat and moisture storage in the ground surface; and vertical transport of latent and sensible heat from the surface. SSTs are computed from specified ocean heat transports and the model's own computations of energy exchange between the atmosphere and the ocean mixed layer.

Linear regression

In order to quantify the SST–precipitation relationship, Sahel rainfall for individual months or seasons (Y) were regressed on monthly or seasonal mean SST at 28 °S, 10 °E (X), the area with the greatest difference in mean SST between the ensembles of five drought versus five rainy simulations in the Druyan (1987) study. In the analysis of these ten model years, when August Sahel precipitation was used as the dependent variable, the linear correlation coefficient of the regression peaked in July at -0.62, indicating that the variations in these SST account for 38 per cent of the inter-annual variance in precipitation. While the correlation with mean March and April SST was almost nil (-0.08 and -0.10, respectively), it strengthened from -0.45 for May SST to -0.51 for June SST. This crescendo, peaking in July and falling off to -0.48 for August SST, lends support to the idea of a causal relationship between the SST and the precipitation, as proposed in Druyan (1987). The influence of the SST thus gradually builds for several months, maximising one month before the event, and then waning as the precipitation regime apparently responds to the already established circulation patterns.

Prescribed change experiments

The objectives of the research were also addressed by designing and analysing a series of prescribed change experiments with the GCM. Various changes in the SST were made on 1 March or on 1 May to a control simulation for which average precipitation was computed for the summer months over the Sahel. These are experiments A1–A7.

An additional aspect of the current study is to determine the sensitivity of the computed drought to initial atmospheric conditions and subsequent atmospheric evolution. The SST anomaly in the South Atlantic that gave the driest summer over north Africa among experiments A1–A7 was tried on a variety of initial atmospheric conditions and different initial SSTs outside the South Atlantic taken from the 1 March model state of previous integrations that serve as controls.

Response to SST patterns

The results of experiments A1–A7 were compared with the simulations from two controls. In each experiment, the atmospheric initial conditions are taken from year 3 of the original 37-year series. The summertime precipitation over the Sahel for this model year was slightly higher than the 37-year mean. The initial SSTs of year 3 are replaced by SSTs taken from year 5, either globally or over the South Atlantic (44–20 °S, 40 °W–20 °E) only, because the Sahel precipitation during the summer of model year 5 was the lowest of the series. An idealised SST anomaly for the South Atlantic was also used to initialise three experimental simulations. Table 14.1 shows the various combinations of initial atmospheric and SST states used for the seven experiments.

June–August precipitation over the Sahel (Table 14.2) and the combined Sahel and Sahara (Table 14.3) was somewhat reduced by all of the experiments relative to the original year 3, although differences were small in A1. Table 14.4 shows that the monthly mean precipitable water over the Sahel was also reduced in every case. A2 shows the greatest impacts due to the SST change on all three tables. Introducing the year 5 South Atlantic SST on 1 March drastically reduces the summer precipitation over most of north Africa (Figure 14.1) as well as the precipitable water. The effect is even more extreme than in the original year 5 from

Table 14.1   Initial conditions for each of seven computational experiments

Experiment	Initial SST	Date of initial conditions
A1	Year 5, global array	1 March
A2	Year 5, S. Atl, rest year 3	1 March
A3	Year 5, global array	1 May
A4	Year 5, S. Atl, rest year 3	1 May
A5	Idealised anomaly S. Atl, rest year 3	1 March
A6	Idealised anomaly S. Atl, rest year 3	1 May
A7	Like A5, but anomaly reintroduced on May 1	1 March

Table 14.2   Sahel rainfall rates (mm/day)

	March	April	May	June	July	Aug	June–Aug	Reduction of year 3 (%)
Orig yr 3	1.7	3.1	2.0	4.0	5.0	5.5	4.8	
Orig yr 5	0.2	1.6	2.7	1.5	3.5	4.1	3.0	38
A1	2.0	2.4	2.2	3.2	5.2	5.5	4.6	4
A2	2.2	1.4	1.1	2.2	4.2	2.9	3.1	35
A3	–	–	2.2	2.7	3.9	4.7	3.8	21
A4	–	–	1.9	3.1	4.8	4.4	4.1	15
A5	1.8	1.7	1.4	2.9	4.3	4.8	4.0	17
A6	–	–	1.9	3.7	3.1	4.4	3.7	23
A7	–	–	1.5	2.4	4.0	4.9	3.8	21

Table 14.3   Combined Sahel and Sahara (12°N–20°N, 10°W–30°E) rainfall (mm/day)

	March	April	May	June	July	Aug	June–Aug	Reduction of year 3 (%)
Orig yr 3	1.0	1.9	1.2	2.0	3.6	4.1	3.2	
Orig yr 5	0.3	0.8	1.2	1.0	2.6	2.9	2.2	31
A1	1.4	1.5	1.1	1.5	3.1	3.6	2.7	16
A2	1.4	1.0	1.0	1.0	2.2	2.3	1.8	44
A3	–	–	1.6	1.2	2.5	3.8	2.3	28
A4	–	–	1.2	1.5	2.5	3.9	2.6	19
A5	1.2	1.0	0.8	1.5	2.4	3.3	2.4	25
A6	–	–	1.1	1.9	2.1	2.7	2.2	31
A7	–	–	0.9	1.4	2.4	3.3	2.4	25

which the SSTs were taken. A6 shows the second largest impacts on the north Africa precipitation.

Results indicate that introducing the prescribed SSTs in March, rather than in May, is not consistently more effective in reducing the subsequent Sahel rainfall: the values shown in Tables 14.2 and 14.3 are generally lower for A3 and A6 than for A1 and A5. The year 5 South Atlantic SSTs are also not consistently effective in exacerbating the simulated drought: tabulated values for A4 do not reflect large

Table 14.4 Sahel precipitable water (mm)

	March	April	May	June	July	Aug	June–Aug	Reduction of year 3 (%)
Orig yr 3	27.3	32.1	31.6	38.3	38.4	40.9	39.2	
Orig yr 5	19.3	27.6	33.6	32.5	36.8	39.5	36.3	7
A1	26.9	28.3	31.7	33.7	37.8	38.6	36.7	6
A2	28.0	23.6	25.9	33.1	35.8	33.8	34.2	13
A3	–	–	33.0	35.6	35.9	38.3	36.6	7
A4	–	–	31.3	34.8	38.0	38.0	37.0	6
A5	26.3	27.5	24.0	31.8	37.2	37.3	35.4	10
A6	–	–	30.8	37.1	35.0	36.3	36.1	8
A7	–	–	21.8	32.5	35.4	36.3	34.7	11

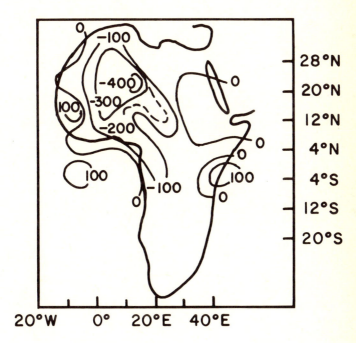

Figure 14.1 Precipitation differences (mm), June to August, A2 – Year 3

impact. Moreover, the idealised SST anomalies showed greater impacts than the year 5 SSTs, with the exception of A2.

The circulation patterns that promote drought in the Sahel evolve from non-linear interactions that are only partially influenced by energy fluxes from the oceans. Still higher SSTs in the south-east Atlantic Ocean brought about reductions in the precipitation and atmospheric moisture over north Africa in all seven experiments. The variations in the impact derive from the influence of each SST pattern and the atmospheric conditions. (Four of the experiments had the atmospheric conditions of 1 March in common and the other three had the atmospheric conditions of 1 May in common.) The particular configuration of the SST pattern used in A2 with 1 March atmosphere of year 3 combined to evolve an extremely dry summer over north Africa. The next section discusses the development of the African summer monsoon as a response to these same SSTs in the South Atlantic in combination with other atmospheric specifications.

Atmospheric initial conditions

The experimental SSTs prescribed in A2 were combined with 1 March initial conditions from 11 other model years from the original 37-year climate history. Table 14.5 summarises the results for the summer months. The computed precipitation for the Sahel region was reduced in eight of the 12 experiments relative to the controls. In each of the eight control simulations, Sahel summer precipitation totals were greater than the 37-year mean. In summary, precipitation was reduced by an average of 25 per cent by the introduction of these SSTs in the south-east Atlantic for their eight experiments. On the other hand, four control simulations with below (3) or near average (1) Sahel summer rainfall became slightly more rainy (an average of 13 per cent) by the introduction of the same SSTs.

Table 14.5 Sahel precipitation (mm), June–August, and reductions compared with each control

	Experiment	Control	Reduction (%)
Year 3	285	442	−36
Year 5	–	285	–
Year 7	335	495	−32
Year 11	303	494	−39
Year 14	359	487	−26
Year 16	430	375	+15
Year 19	340	533	−36
Year 22	413	524	−21
Year 23	400	366	+9
Year 24	430	366	+17
Year 28	480	427	+12
Year 30	377	460	−18
Year 31	450	515	−13
Average reduction			−14

Ensemble means were made for the five experiments that showed the greatest reduction in Sahel summer rainfall due to the experimental SST (ENS1). Figure 14.2 shows the initial difference in SST between the experimental values and the ensemble means from the controls on 1 March. The prescribed change in SST represents a coherent warm anomaly in the South Atlantic Ocean whose maximum is 2.2 °C and a cold anomaly of the same intensity to the west. This pattern includes a slightly more restricted area of positive SST difference that was evident between the initial SSTs of year 5 and year 3 discussed above. The SST fields evolve according to the interactive ocean formulation and the mean March SST differences (Figure 14.3) show a slight moderation of the extrema of the SST differences as

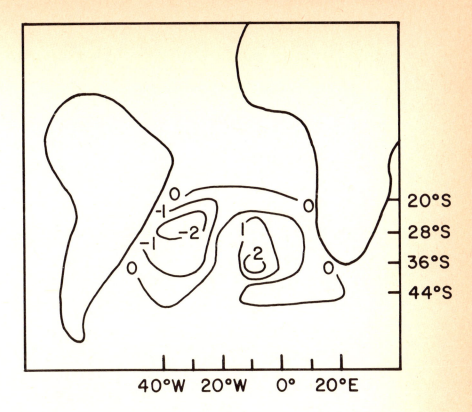

Figure 14.2 SST differences (°C), 1 March, Year 5 − Year 3

compared to the initial conditions. These extrema are smaller yet in April although the same general pattern appears. By May the SST excess has decreased to below 1 °C in the area of the original maximum difference and the negative difference closer to South America continues to wane. By June, the only noteworthy change is the strengthening of the SST excess to 1.5 °C in the central South Atlantic. This region shows additional relative warming through July and August while maintaining a more modest excess of 0.5–1.0 °C in the south-east Atlantic closer to South Africa. This last feature is similar to the SST excess of the dry ensemble discussed in Druyan (1987), and it is consistent with the observational studies that found warm SSTs in the south-east Atlantic associated with dry summers in the Sahel (Lamb 1978; Lough 1986; Folland *et al.* 1986). Druyan (1987) found that these warm SSTs cause lower sea-level pressure (SLP) that weaken the southerly flow advecting moisture to north Africa.

Inspection of the monthly mean latitudinal component of the surface wind shows lower values in ENS1 adjacent to south-west Africa in March–July than in its control ensemble. These are, of course, indicative of the weaker southerly flow over the south-east South Atlantic, a characteristic also present in the dry versus wet ensembles of Druyan (1987).

Another set of ensemble means (ENS2) was made from three of the four

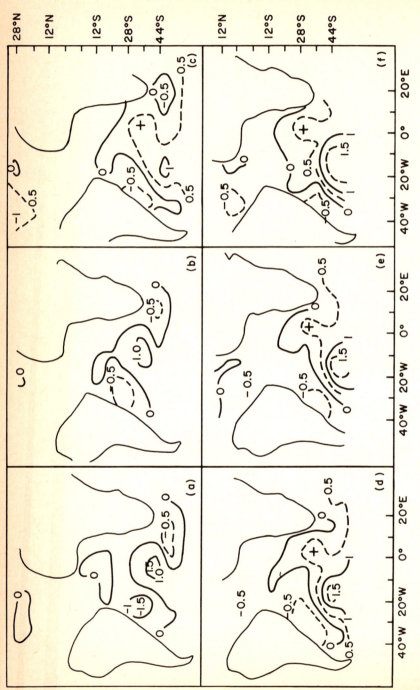

Figure 14.3 SST differences (°C) for ensemble monthly means, ENS1 − Control: (a) March, (b) April, (c) May, (d) June, (e) July, (f) August

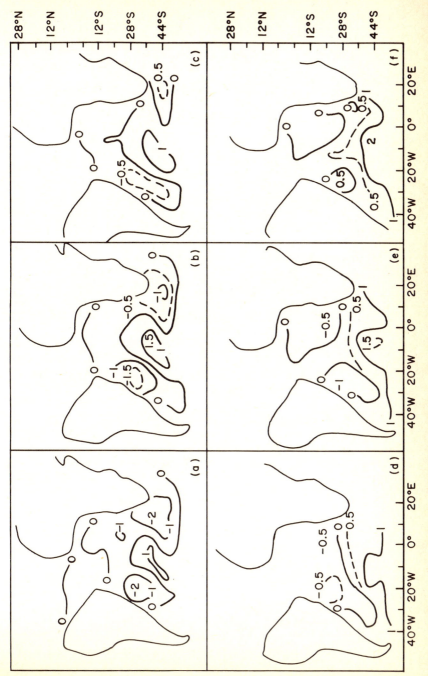

Figure 14.4 As Figure 14.3, but ENS2 — Control

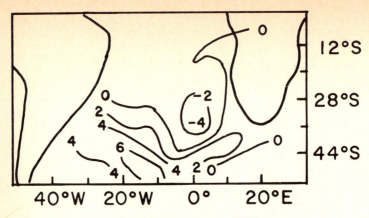

Figure 14.5 SLP differences (mb), May, ENS1 − ENS2

simulations that experienced increases in Sahel rainfall despite initialisation on 1 March with the same SSTs in the South Atlantic as in ENS1. (Data for the fourth were lost by a computer failure.) The average increase in June–August Sahel precipitation for these three simulations was a modest 14 per cent compared to the −33 per cent impact registered by the ENS1 simulations). The SST differences (ENS2−control) developed somewhat differently than in the ENS1 experiments. Relatively warm SSTs in the controls near South Africa on 1 March created a pattern of negative differences in the initial conditions. Figure 14.4a shows a broad area of −1 °C to −2 °C in the south-east South Atlantic between 0° longitude and Africa in March. The positive anomaly area moved closer to the continent in April and May, but it was consistently weaker east of 0° longitude than the ENS1 ensemble SST anomalies. After May, the characteristic tongue of positive SST differences over the south-east South Atlantic was missing completely (north of 28 °S), in further contrast to the SST differences that evolved in the ENS1 experiments (Figure 14.3). In the following discussion, the ensemble means of other ENS1 and ENS2 output variables are compared in order to study the causes for ENS2 not developing the drought conditions over the Sahel that were evidenced in ENS1.

Differences between these ensembles show generally higher rainfall rates for ENS2 over north Africa along 12 °N in June, July and August. As in the previous study, differences in the advection of moisture from the south-east Atlantic were found to be consistent with this divergence in the Sahel precipitation. Rather large and extensive differences in the southerly component of the surface wind that favour the moisture advection in ENS2 from the South Atlantic to north Africa can be discerned for the months of March, May, July and August with less impressive differences in the same sense for April and June (not shown).

These ensemble differences in the strength of the southerlies over the south-east Atlantic imply differences in the gradients of SLP. Figure 14.5 shows, for example, that the ensemble differences in SLP for May reached 2–4 mb along 0° longitude. Indeed, the ENS2 simulations maintained higher pressures over the south-west South Atlantic while the ENS1 ensemble did not develop the ridge in this area until August. Thus, the east–west SLP gradient necessary for vigorous southerlies near 0° longitude evolved in ENS2 and not in ENS1. Higher SLPs do evolve in the

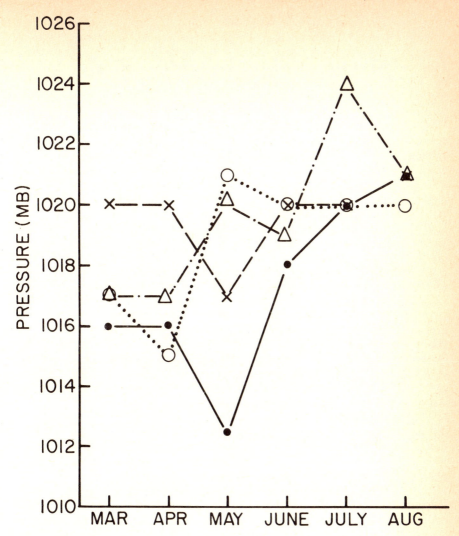

Key:
. . . . ENS1; triangles: control for ENS1
xxxxx ENS2; circles: control for ENS2

Figure 14.6   SLP maxima of South Atlantic ridge (28 °W–44 °S, 0 °N–10 °W)

ENS1 set nearer to South Africa in July and persist into August. However, this high pressure region effectively extends the ridge along 28 °S eastward from 10 °W, preventing the generation of strong southerlies like those simulated by the ENS2 experiments at this latitude.

The SST of much of the South Atlantic cooled more slowly in the ENS2 ensemble between March and April and remained about 0.5 °C warmer from April through August. As discussed above, the ENS2 subtropical ridge is somewhat stronger than the ENS1 ridge, despite the slightly warmer SST. (Higher SLPs at

10 °E are, however, consistent with cooler SST generated there in ENS1.) The difference in the SLP pattern seems to be related to the better definition of the ridge west of 0° longitude in the initial conditions for the ENS2 experiments on 1 March. Thus, the mean March SLP distribution shows a ridge extending from 10 °E to 30 °W along 44 °S, including maxima at 0° and 30. The March SLP of the ENS1, on the other hand, confines the ridge entirely to the eastern side of 0° longitude with troughing along 10 °W.

Figure 14.6 shows the monthly mean SLP maxima (at the ridge line) within the area 28–44 °S, 10 °W–0° for the two experimental ensembles and their control ensembles. Note the dramatic increase in the ridge SLP in both experimental ensembles in May, presumably a delayed reaction to the positive temperature anomalies. Although the time trends are parallel, these SLP maxima of ENS2 are 2–4 mb higher in March–June. Note also that the ensemble with the driest Sahel summer (ENS1) has the weakest ridge.

The ridge maintained in the ENS2 simulations was stronger than its counterpart in the control simulations as early as March. Not only were the ENS2 experiments initialised with atmospheric conditions that included a better-defined subtropical ridge, their prescribed SSTs in the South Atlantic were 1–2 °C cooler than the control SST over a broad area near South Africa (while the warm anomaly was situated farther west; see Figure 14.4). This reinforced the high pressure near 0° longitude. These two influences in favour of higher values of the SLP maxima help explain how a regime conducive to northward moist advection was maintained despite the subsequent weakening of the ridge by warm SST. A more detailed analysis of the atmospheric dynamics that supported the persistence of a relatively strong subtropical high, despite the same initial SST as in ENS1, cannot, however, be gleaned from the monthly mean synoptic fields that are the basis of the present study.

Discussion

The largest consistent response to the prescribed SST anomaly in the experiments was to create drier than usual conditions for the Sahel during the summer months. Since this occurred for a variety of initial atmospheric regimes (that preceded rainier summers in the control simulations), we can interpret the results as a demonstration of SST forcing that overrides the influences of synoptic regimes that may otherwise be favourable to abundant Sahel rainfall. SST anomalies can persist many months, so that an anomaly which first appears in March can indeed have an influence on rainfall during the following summer. A number of experiments also showed that the drought-inducing impact of SST anomalies can be suppressed when the atmospheric dynamics force the evolution in a direction favourable to greater humidities over north Africa. In particular, a strong subtropical ridge over the south-east South Atlantic in the spring can mitigate the influence of warm SST anomalies which can otherwise establish a synoptic regime that leads to drought over north Africa.

Acknowledgements

The data processing for the research was efficiently and graciously accomplished by the efforts of Dr R Ruedy. Dr David Rind was a constant consultant during the research and the writing of the manuscript. J. Mendoza drafted the figures.

References

Druyan, L.M. 1987. 'GCM studies of the African summer monsoon', *Climate Dynamics* 2, 117–26.

Folland, C.K., Palmer, T.N., Parker, D.E. 1986. 'Sahel rainfall and worldwide sea temperatures, 1901–85', *Nature* 320, 602–7.

Hansen, J., Russel, G., Rind, D., Stone, P., Lacis, A., Lebedeff, S., Reudy, R., Travis, L. 1983. 'Efficient three-dimensional global models for climate studies: models I and II', *Mon. Wea. Rev.* 111, 609–62.

Lamb, P.J. 1978. 'Large-scale tropical Atlantic surface circulation patterns associated with sub-Saharan weather anomalies', *Tellus* 30, 240–51.

Lough, J.M. 1986. 'Tropical Atlantic sea surface temperatures and rainfall variations in Subsaharan Africa', *Mon. Wea. Rev.* 114, 561–70.

Chapter 15

Sea-surface temperature anomaly patterns and prediction of seasonal rainfall in the Sahel region of Africa

D.E. Parker, C.K. Folland and M.N. Ward Meteorological Office, United Kingdom

Introduction

Analysis of the Meteorological Office Historical Sea Surface Temperature (MOHSST) data set (Parker and Folland, Chapter 4, this volume) has demonstrated that significant very large-scale fluctuations of sea-surface temperature (SST) have occurred during the present century on time-scales from months to many decades. This century has also witnessed marked fluctuations of rainfall in sub-Saharan Africa, not only on the inter-annual time-scale, but also on decadal scales, especially the dramatic decline in rainfall since the 1950s (Nicholson 1985; see also Figure 15.1). The aim of this paper is to show that statistical relationships between Sahel rainfall and recurring patterns of anomalous world-wide SST (Folland, Palmer and Parker 1986) are physically well founded, and to describe two statistical techniques for prediction of seasonal Sahel rainfall, illustrating their suitability when tested on independent data.

World-wide sea-surface temperature and Sahel rainfall

Figure 15.2 taken from Folland, Palmer and Parker (1986), shows the difference in July to September SST between the five driest and five wettest Sahel rainfall seasons of the period 1950–86. Most of the seasonal rainfall in the Sahel falls in July to September. The pattern of SST differences mainly reflects an inter-hemispheric scale of changes, but with the northern Indian Ocean changing in phase with the southern hemisphere, and with very little signal in the south-west Pacific. A warmer southern hemisphere and Indian Ocean corresponds to drier conditions in the Sahel. An apparent indication in Figure 15.2 that 'El Niño' events (warm SST in the tropical east Pacific) are associated with dry Sahel is, however, not quite statistically significant, and partly results from the coincidence of the extreme El Niño of 1982 and 1983 with very dry conditions in the Sahel in both years. In 1984, when the Sahel was actually at its driest (Figure 15.1), the topical east Pacific was much colder than in 1982 and 1983.

We can estimate the likely magnitude of the effects of world-wide SST anomalies on the atmosphere by using the simple physical arguments of Sawyer (1965). For example, imagine a world-wide pattern of SST changes to be suddenly 'switched on', and (as in Figure 15.2) that the regional scales of the changes are approximately the same as those of the largest scale features of the atmosphere. Let the mean modulus of the changes be 0.46 °C to correspond with Figure 15.2.

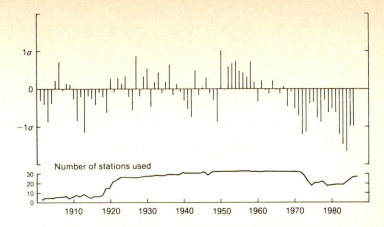

Figure 15.1 Standardised annual rainfall anomalies for the Sahel, 1901–86

Sources: Values to 1984 after Nicholson; 1985–86 estimated from CLIMAT reports.

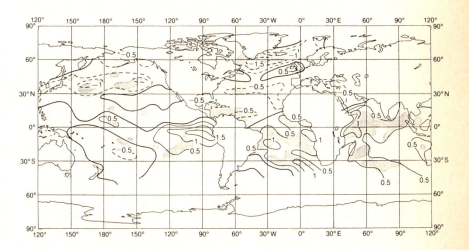

Figure 15.2 Sea-surface temperature, July to September: average of (1972–3, 1982–4) (Sahel dry) minus average of (1950, 1952–4, 1958) (Sahel wet)

Note: Contours every 0.5 °C. Shaded areas are different from zero at the 90% level according to a *t*-test.

Assuming that the atmosphere is initially in its 'climatological mean state', this magnitude of SST changes would generate anomalous heating on these atmospheric scales with a mean modulus of about 15 W m^{-2} over the 70 per cent of the globe that is ocean. The local magnitude of the heating change would vary from zero to substantially larger values. This figure of 15 W m^{-2} is about 15 per cent of the maximum change in the large-scale heating of the atmosphere by the oceans during the seasonal cycle. So world-wide changes of large-scale SST pattern of the magnitude typical of Figure 15.2 (which is a smooth difference pattern based on

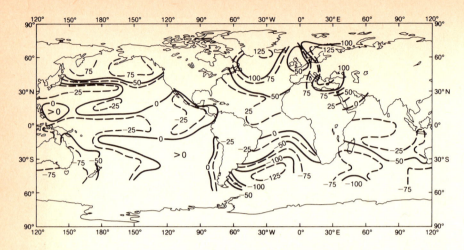

Figure 15.3   Third covariance eigenvector of worldwide seasonal SST anomaly variations, 1901–80. (EOF weightings have been multiplied by 100.)

two sets of non-consecutive five-year periods, one set very dry and the other very wet in the Sahel), are potentially large enough significantly to alter the mean global atmospheric circulation.

For the SST anomaly pattern in Figure 15.2 to be associated confidently with Sahel rainfall fluctuations over a longer period, the near inter-hemispheric contrast in SST anomalies must be a principal and recurring pattern of SST variability. Encouragingly, an eigenvector or 'empirical orthogonal function' (EOF) analysis of world-wide SST anomalies (on a $10° \times 10°$ space scale and seasonal time-scale, and with all seasons included in the analysis) for the period 1901–80 reproduces, as the third EOF (see Figure 15.3), the essence of the north–south contrast in Figure 15.2. The data on which Figures 15.2 and 15.3 are based are largely, but not entirely, independent. However, analysis of decadal SST anomaly fields confirms that inter-hemispheric variability has been present for much of this century. For example, the strong inter-hemispheric contrast contained in the 1968–86 SST anomaly field (relative to 1951–80) (Figure 15.4a) was also prominent in the 1911–20 field (relative to 1901–30) (Figure 15.4), though details in the equatorial Atlantic do seem to have fluctuated. We have chosen periods for the climatological averages to be contemporary with the decades considered so as to remove the considerable globally averaged SST changes which do not seem to be related to Sahel rainfall variations.

The near hemispheric pattern of SST variability is therefore believed to be real and to constitute one component of an interrelated set of changes in the global climate system. One consequence of those changes seems to be variability in Sahel rainfall.

The time coefficients of an EOF pattern trace the contribution of that pattern to observed SST anomaly fields. The time coefficient of the inter-hemispheric pattern EOF 3 in July to September is very significantly correlated with Sahel rainfall over the period 1901–85 (see Figure 15.5, $r = +0.62$ (significance = 0.1% assuming only 32 degrees of freedom because of autocorrelation)). Most importantly, the correlation is quite stable (1901–45, $r = +0.50$ (significance = 1% assuming 24 degrees of freedom) and 1946–85, $r = +0.71$ (significance = 5% assuming 8 degrees of freedom)), even though the nature of the variability of Sahel rainfall has changed, being predominantly inter-annual in the earlier part of the century and interdecadal in the later part. The earlier SST data are, of course, less complete.

Some of the inter-annual and inter-decadal variability in Sahel rainfall which is unaccounted for by the interhemispheric pattern of SST anomalies may be linked to other principal patterns of SST variability. Indeed, Sahel rainfall was found to be correlated with EOF 6 ($r = +0.44$ for 1901–85, significance = 1%), which mainly represents a contrast between North and South Atlantic SST anomalies, especially in the Benguela current region (Figure 15.6). The Indian Ocean anomalies are in phase with those of the South Atlantic, and much of the Pacific Ocean is in phase with the North Atlantic. The sign of EOF 6 is again chosen to be that having a positive correlation with Sahel rainfall. However this relationship is less clear than that with EOF 3 (Figure 15.7) and appears to apply only during periods when variations of Sahel rainfall have been predominantly inter-decadal (see correlations marked on Figure 15.7). This observation supports Lough (1986), who found that significant relationships established for 1948–72 between Sahel rainfall and a tropical Atlantic SST anomaly pattern quite similar to EOF 6 did not exist in 1911–39.

The correlation between Sahel rainfall and the July to September coefficient of EOF 2, which represents a global signal of El Niño in SST anomalies, was only around -0.2 (El Niño very weakly associated with a dry Sahel) for both 1901–45 and 1946–85. El Niño is probably only an appreciable influence when it is strong during the rainfall season; Palmer (1986) provides physical evidence for an influence of a strong El Niño SST pattern on Sahel rainfall.

A final argument to support the physical reality of the SST–Sahel rainfall relationships is the success of recently improved atmospheric general circulation models in reproducing the observed changes in rainfall and atmospheric circulation when the models are forced by appropriate SST changes. For earlier work the reader is referred to Folland, Palmer and Parker (1986) and to Palmer (1986). The chapter by Owen and Folland in this volume (Chapter 13) presents the most comprehensive results so far.

Prediction of Sahel rainfall

We have developed two statistical methods to predict Sahel seasonal rainfall from SST anomalies. Both methods depend on the fact that relationships very similar to those described above were found to hold between Sahel rainfall and SST anomalies observed several months in advance.

The first, more elementary, method produces a forecast by computing the similarity between the pattern of SST anomalies measured prior to the main Sahel rainfall season and a pattern of linear regression slopes calculated by fitting quotations like that below over a historical period:

$$x_i = m_i R + d$$

where R is Sahel rainfall, x_i is the SST anomaly observed prior to the main rainfall season in the ith $5° \times 5°$ grid square, m_i is the corresponding regression slope, and d is a constant. A substantial historical or 'training' period such as 1946–84

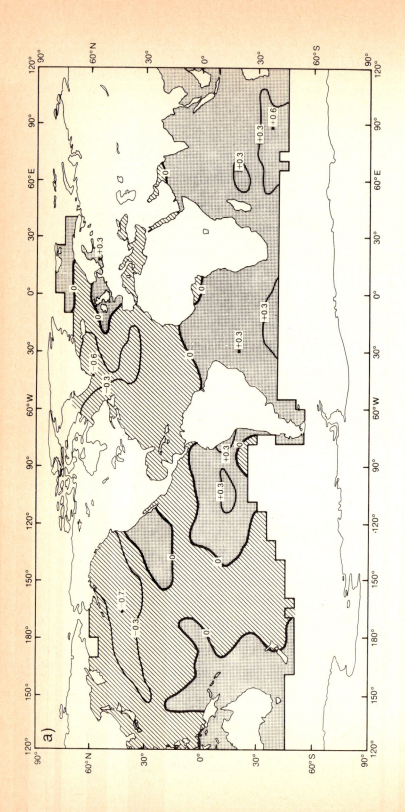

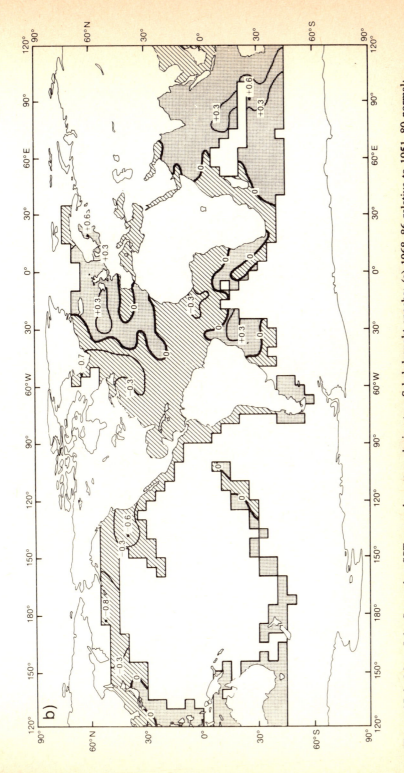

Figure 15.4 Mean July–September SST anomaly pattern during two Sahel drought epochs: (a) 1968–86 relative to 1951–80 normal; (b) 1911–20 relative to 1901–30 normal

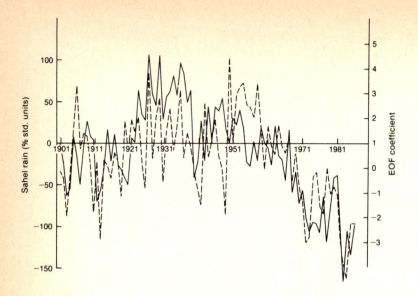

Key:
——— Coefficient of world-wide SST E of 3 in July to September
-------- Sahel rainfall, 1901–86

Figure 15.5   Coefficient of world wide SST EOF 3 in July to September, along with Sahel rainfall, 1901–86

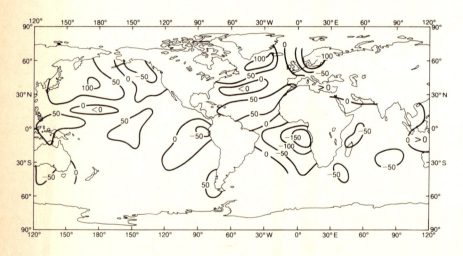

Figure 15.6   Sixth covariance eigenvector of world wide seasonal SST anomaly variations, 1901–80

Note: EOF weightings have been multiplied by 100.

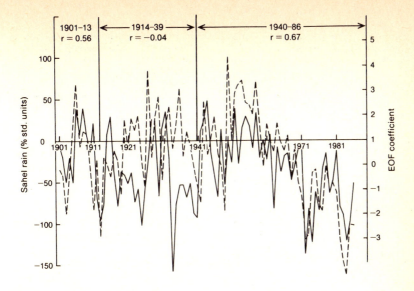

Key:

——— Coefficient of worldwide SST EOF 6 in July to September
-------- Sahel rainfall, 1901–86

Figure 15.7   Coefficient of worldwide SST EOF 6 in July to September, along with Sahel rainfall, 1901–86

was used to determine the values of m_i. The fields of linear regression slope contain positive and negative values whose pattern is rather similar to a combination of Figures 15.2 and 15.3. After first subtracting the global mean SST anomaly from each $5° \times 5°$ anomaly, the covariance C was then calculated between the regression slope field and the $5° \times 5°$ SST anomaly field in the same calendar months for each year in the training period. A second linear regression of subsequently observed Sahel rainfall (R) on C was then carried out for the training period to provide a prediction equation. C is a measure of similarity between the SST field in a given year and the field of values of m_i corrected for global mean climatic changes in SST. Values of C were also computed for independent years, e.g. 1901–39 or 1985–7, so that the linear regression of R on C could be used to 'predict' Sahel rainfall.

The second method uses multivariate linear discriminant prediction, a technique also used to make experimental monthly long-range forecasts for the United Kingdom (Folland and Woodcock 1986). It is more complex, but allows more than one degree of freedom to be distinguished in the large-scale fields of SST anomalies that may influence Sahel rainfall. In the version described here, predictors are selected from coefficients of EOFs of the global fields of SST anomaly measured several months in advance of the Sahel rainfall season, again using a substantial training period such as 1946–84. The EOFs are predetermined patterns based on SST for 1901–80 (as in Figures 15,3 and 15.6) or 1951–80. Selection of predictors is made using a stepwise technique employing multivariate analysis of covariance together with other tests on the dependent data that attempt to make a priori

Table 15.1   Summary of some experimental forecasts

Training period	Forecast period	Persistence of Sahel rainfall in forecast period (auto-correlation with one year's lag)	Months used for forecast	Regression technique correlation of forecast and observed values	Discriminant prediction (the predictors were the coefficients of 1901–80 EOFs 3 and 6)	
					Correlation of: forecast category and observed category*	RPS score (%)
1901–45	1946–86	0.69	March & April	0.26	0.73	24.1
			April & May	0.57	0.71	28.0
			May & June	0.77	0.31	18.7
1946–86	1901–45	0.11	March & April	0.12	0.17	23.4
			April & May	0.30	0.34	22.7
			May & June	0.46	0.27	24.5

* Categories are measured on a scale of 1 (very dry) to 5 (very wet)

estimates of predictive skill. The quantities predicted are the probabilities that the Sahel rainfall will be in each of five climatologically equiprobable ranges or 'quints' defined using the training period. As expected, the coefficients of 1901–80 global EOFs 3 and 6 (Figures 15.5 and 15.7) have been found to dominate the predictors while global mean changes in SST (EOF1) have no significant skill. In sets of discriminant prediction equations ('models') which use the SST anomaly field in February and March to predict Sahel rainfall, inclusion of the El Niño (1901–80 EOF2) pattern also has a beneficial impact on forecasting skill. When SST anomalies closer to the rainfall season are used, the additional skill that results from including a direct measure of El Niño disappears, as would be expected from the lack of significance in Figure 15.2 and the low simultaneous correlation between Sahel rainfall and the coefficient of (1901–80) EOF2. The reason for this decline in influence as the Sahel rainfall season is approached may be that information from the mature phase of El Niño has been transmitted, through ocean–atmosphere interaction, into the quasi-hemispheric SST anomaly fields especially in the Atlantic and Indian Oceans (Pan and Oort 1983; Wolter 1987), thus changing the coefficients of 1901–80 EOFs 3 and 6.

Table 15.1 summarises the skill attained by the two statistical methods in forecasting (or hindcasting) Sahel rainfall during periods independent of training data. The Epstein ranked probability score (RPS) (Daan 1985) that has been used to assess the probability forecasts provided by the discriminant prediction technique has a chance value of zero and a maximum of 100%. Both forecasting methods, and especially linear discriminant prediction, show substantial skill in each forecast period despite the radically different statistical nature of the Sahel rainfall series in the two periods. An unexplained feature is the poorer performance of discriminant prediction in forecasting 1946–86 rainfall using May and June SST from a 1901–45 training period. Even the better set of forecasts of 1946–86 rainfall using April to May SST had a tendency to predict quint 2 (dry) when quint 1 (very dry) was observed (Table 15.2a). The reason may lie in the small number of truly very dry cases in the 1901–45 training period (Table 15.2c), resulting in an inability to

Table 15.2 Relationships between Sahel rainfall forecasts and observed values by linear discriminant prediction

(a) Forecasts for 1946–86 using model with 1901–45 training period, in terms of quint boundaries. Averaged April to May SST anomaly fields form the EOF coefficient predictors.

Observed quint:		1	2	3	4	5
Nicholson's (1985) standardised units:		< −42	−42 to −21	−20 to −2	−1 to +27	>27
	1	3	0	2	0	0
Forecast	2	11	3	1	1	0
quint	3	1	0	1	1	0
	4	1	2	1	2	5
	5	0	0	1	1	4

(b) As (a) but 45-year-long training periods and quint boundaries updated every five years.

Observed quint:		1	2	3	4	5
	1	14	3	1	0	0
Forecast	2	2	0	2	2	0
quint	3	1	0	0	0	0
	4	1	1	2	0	2
	5	1	1	2	1	5

(c) As (a) but forecasts for 1901–45 using 1946–86 training period and quint boundaries.

Observed quint:		1	2	3	4	5
Nicholson's (1985) standardised units:		< −87	−87 to −44	to −43 −11	to −10 +24	to <24
	1	0	0	0	0	0
Forecast	2	1	2	2	0	0
quint	3	0	1	3	3	1
	4	1	2	9	6	7
	5	0	1	2	2	2

properly distinguish between quint 1 and quint 2 in the training data because their mean values of rainfall are fairly close in that period. When the training period was progressively updated so as to end within five years of the year to be forecast (the quints were progressively redefined to be those appropriate to the given training period), this fault was rectified (Table 15.2b) because the difference in mean rainfall between quints 1 and 2 and also in their associated sea-surface temperature patterns became progressively larger.

Figure 15.8 shows that the skill of the regression method in hindcasting 1901–45 Sahel rainfall from May to June SST using 1946–86 training data included successful hindcasts of the very dry years 1913 and 1903. However, hindcasts of wet years are poor, and there is a clear bias towards forecasts of low rainfall.

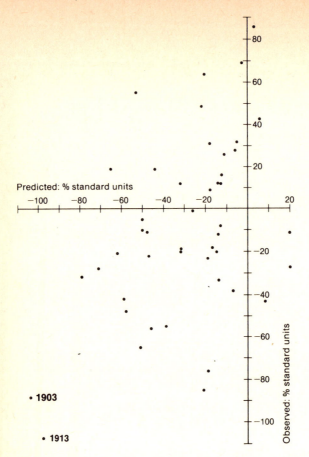

Figure 15.8 Regression technique Sahel rainfall predictions for 1901–45 using the average of May and June SSTs and a 1946–86 model training period

Note: Forecasts for the two driest years (1913 and 1903) are specifically indicated. See text for a brief discussion of bias.

Forecasts for 1986 and 1987

By June 1986 the work reported here was sufficiently developed to justify the issuing, by the Meteorological Office, of a cautiously worded experimental forecast for Sahel rainfall for the summer of 1986. The regression method indicated that rainfall would be about 71 pr cent of the 1951–80 average[1] rainfall; the more complex discriminant analysis prediction is shown in Table 15.3 and favoured quint 1 and quint 2 with a near equal probability. Based on the currently available CLIMAT message data, observed rainfall was about 66 per cent of the 1951–80 normal.

For 1987, a forecast was issued in May and subsequently updated in June. The regression method predicted 69 per cent of the 1951–80 normal rainfall; discriminant analysis predictions are in Table 15.3, this time providing a clear bias toward quint 1 (very dry). Indications to date (late August 1987) are that Sahel rainfall has

Table 15.3   Issued discriminant prediction technique probability forecasts for Sahel
rainfall categories: (a) forecast for 1986; (b) forecasts for 1987.

(a)

Quint (1946–84)	1	2	3	4	5
% 1951–80 normal*	<77	77–93	94–104	105–119	>119
1986 probability forecast from April and May SSTs	0.32	0.34	0.26	0.07	0.01

(b)

Quint (1941–85)	1	2	3	4	5
% 1951–80 normal*	<75	75–91	92–102	103–117	>117
1987 probability forecast from March and April SSTs	0.60	0.20	0.12	0.07	0.01
1987 probability forecast from April and May SSTs	0.72	0.18	0.06	0.02	0.01

* Approximate conversion from Nicholson's (1985) standardised rainfall units for stations used.
Note: The quint categories in (a) and (b) are equiprobable over slightly different training periods,
hence the quint boundaries differ slightly.

again been substantially below normal, but full verification, of course, awaits the
end of the 1987 rainfall season.

Conclusion

We believe that a useful start in understanding and forecasting Sahel rainfall has
been made. Research continues towards establishing separate forecasting equations
for the eastern and western parts of both the Sahel and the Soudan, which is the
climatologically moister belt immediately to the south (Nicholson 1985). The
eastern parts of both regions are probably more influenced by the Indian Ocean
(Palmer 1986). More fundamentally, we plan a series of extensive atmospheric
general circulation model integrations forced with observed SST and perhaps polar
ice limits for a number of complete cycles to explore whether it might be possible
to forecast seasonal and perhaps intra-seasonal tropical rainfall in a more complete
way.

Note

1. The 1986 issued forecasts were expressed relative to a 1964–84 climatology. Here we
   use 1951–80 normals for consistency with other work.

References

Daan, H. 1985. 'Sensitivity of verification scores to the classification of the predictand', *Mon. Wea. Rev.* 113, 1384–92.

Folland, C.K., Palmer, T.N., Parker, D.E. 1986. 'Sahel rainfall and worldwide sea temperatures 1901–85', *Nature* 320, 602–7.

Folland, C.K., Woodcock, A. 1986. 'Experimental monthly long-range forecasts for the United Kingdom. Part I. Description of the forecasting system', *Met. Mag.* 115, 301–18.

Lough, J.M. 1986. 'Tropical Atlantic sea surface temperatures and rainfall variations in sub-Saharan Africa', *Mon. Wea. Rev.* 114, 561–70.

Nicholson, S.E. 1985. 'Sub-Saharan rainfall 1981–84', *J. Clim. Appl. Met.* 24, 1388–91.

Palmer, T.N. 1986. 'Influence of the Atlantic, Pacific and Indian Oceans on Sahel rainfall', *Nature* 322, 251–3.

Pan, Y.H., Oort, A.H. 1983. 'Global climate variations connected with sea surface temperature anomalies in the eastern equatorial Pacific Ocean for the 1958–73 period', *Mon. Wea. Rev.* 111, 1244–58.

Sawyer, J.S. 1965. 'Notes on the possible physical causes of long-term weather anomalies', *WMO-IUGG Symp. Research and Development Aspects of Long-Range Forecasting, Boulder, Colorado, 1964*, WMO no. 162, TP79, Tech. Note 66, 227–48.

Wolter, K. 1987. 'The Southern Oscillation in surface circulation and climate over the tropical Atlantic, eastern Pacific and Indian Oceans as captured by cluster analysis', *J. Clim. Appl. Met.* 26, 540–58.

Chapter 16

Changes in wet season structure in central Sudan, 1900–86

M. Hulme   University of Salford, United Kingdom

Introduction

Since the mid-1960s the existence and character of climatic change in the African Sahel has stimulated extensive debate among climatologists. Many workers have investigated annual rainfall series (e.g., Bunting *et al.* 1976; Nicholson 1979; Lamb 1983) and, more recently, others have investigated monthly rainfall series (e.g., Gregory 1983; Nicholson and Chervin 1983; Dennett *et al.* 1985). There are two gaps in this research literature. First, the analyses have concentrated largely on the west African Sahel, largely ignoring the area east of 20 °E, i.e. Sudan and Ethiopia. Sudan, however, is part of a climatically contiguous belt stretching 5,000 km across the African continent. This chapter contributes a perspective from the eastern Sahel by analysing data from central Sudan between 12° and 16 °N (Figure 16.1). Second, most studies have used annual and monthly rainfall totals and largely neglected daily rainfall series. In Sudan, the only investigations based on daily falls have been for individual years or small samples of years (e.g., Hammer 1968; Pedgley 1969). This omission is unfortunate, because analysis of daily rainfalls as opposed to annual rainfalls reveals more clearly the potential impacts of climatic change on human society and the physical landscape.

This chapter uses long-term series of daily rainfalls from central Sudan to investigate secular changes in the distribution of rainfall through the wet season (hereafter termed 'wet season structure'). A large literature is devoted to one particular aspect of this problem, namely the date and variability of the onset of the rains in the monsoonal climates of west Africa, India and north Australia. This chapter, however, is concerned with slightly wider issues. Have there been any systematic changes in the onset, duration or termination of the wet season in central Sudan? If so, have these coincided with the well-established periodic variations in Sahelian annual rainfall totals noted elsewhere (e.g., Nicholson 1979; Lamb 1983; Trilsbach and Hulme 1984)? Finally, has the frequency of damaging dry spells within the wet season, or complete failures of the wet season, changed significantly?

Such 'secondary parameters' have been recommended when examining rainfall changes in the Sahel (Trilsbach and Hulme 1984). They may be of greater utility for increasing the compatibility between models of climate impact and models of climatic change (Parry and Carter 1984) than would crude annual totals, the 'primary parameter'. Not only do they relate to the seeding, tending and harvesting schedules of rainfed crops (Benoit 1977), they also affect the migratory regimes of large numbers of pastoralists, the opening and closing of mechanical boreholes for water supply (Hulme 1986) and the run-off–recharge ratio of precipitation.

After first commenting on the range of subtropical wet season definitions, a suitable wet season model is constructed and applied to 12 long-term series (> 60

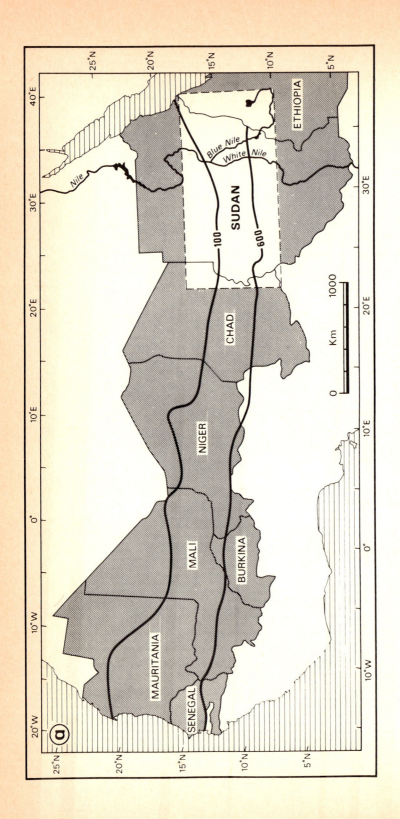

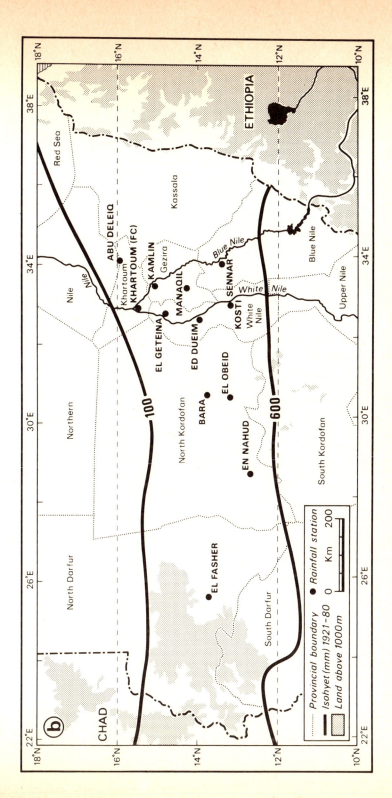

Figure 16.1 Location of (a) Sahelian states, and (b) 12 central Sudan rainfall stations used in the full series analysis

years) of daily rainfalls in central Sudan. The parameters of the wet season structure (onset, duration, termination, breaks) are then examined for changes during the twentieth century. Two years are then selected for further study, 1983 and 1986. The model is applied to over 30 stations for these years and the spatial pattern of the duration of the wet season is examined. A concluding discussion addresses the value of such a model for climatic change studies, agricultural and vegetation monitoring.

Wet season definitions for the subtropics

There are two broad categories of subtropical wet season definition: those formulated in terms of absolute daily or monthly rainfall totals and those formulated in relation to some standardising parameter (e.g., the magnitude of annual rainfall or evapotranspiration). These two categories can be termed 'absolute' and 'relative' respectively. Relative definitions have the advantage of being applicable to a wider range of climatic regimes than can absolute definitions. Both categories can be subdivided according to the scale of measurement used: months, pentads or individual days. Further distinctions exist between those definitions that are applicable on a year-by-year basis rather than to, say, 30-year mean values only and also between those that are based on rainfall parameters alone rather than a range of climatic parameters.

The adoption of a wet season definition and model

For the present study, the definition of the wet season adopted: (1) does not vary with annual rainfall totals; (2) is relative and is standardised by potential evapotranspiration; (3) is calculable on a year-to-year basis; (4) utilises the most detailed scale of data available (i.e. daily rainfall data); and (5) detects dry spells within the wet season and makes allowance for 'false starts'. The first four criteria have been mentioned above. It is important to include the fifth because of the agricultural implications of an applied wet season definition. Where the wet season is short, as in central Sudan, the cultivator is always tempted to initiate seeding at the first major rainfall of the year (Agnew 1982). If this proves to be an isolated storm, with the bulk of the rains still pending, germination of seeds is unlikely to occur. Any definition of the onset of the wet season should exclude such 'false starts' to prevent spuriously long wet seasons. Similarly, the occurrence of dry spells within the heart of the wet season can be damaging for plant growth and almost certainly leads to yield reductions (Benoit 1977).

A simple water balance model was therefore developed to allow a detailed examination of changes in the structure of the wet season, rather than to monitor precisely hydrological conditions in the surface soil. The model consisted of daily rainfall and daily potential evapotranspiration (ET_p) with the daily balance determined as follows:

$$w_{i+1} = w_i + r_{i+1} - f(ET_p) \qquad (1)$$

where w_i is the water balance of day i (> 0), r_{i+1} is the rainfall of day $i + 1$, and ET_p is potential evapotranspiration (based on Penman).

Run-off and deep throughflow were initially set at zero and soil moisture capacity as infinite. Surface discharge and significant throughflow in the sandy soils of central Sudan are both minimal because of low relief, high infiltration and rapid evaporation from the subsurface layers. In view of criterion (5) above, the following general formulae were established for defining the onset and termination of the wet season and breaks within the wet season:

$$o_j = \text{smallest } i \text{ where } w_i > 0 \text{ and } w_i \text{ to } w_{i+x} > 0 \qquad (2)$$

$$t_j = \text{largest } i \text{ where } w_i > 0 \text{ and } w_{i-x} \text{ to } w_i > 0 \qquad (3)$$

$$l_j = t_j - o_j \qquad (4)$$

$$b_j = \text{a sequence of} > \text{days where } w_i \text{ to } w_{i+y} = 0 \qquad (5)$$

where o_j is the onset date in year j, w_i the water balance on day i (i = 1, 365) as defined in Eq. (1), t_j the termination date in year j, l_j the wet season duration in year j, and b_j a break within wet season in year j. The model was then run several times varying the values of x and y in Eqs (2), (3) and (5), with the function of ET_p (crop coefficient) in Eq. (1) also being varied between 0.5 and 1.0. From these trial runs the following values were selected: $x = 10$, $y = 15$ and $f(ET_p) = 0.5(ET_p)$.

A minimum ten-day period of positive water balance to open and terminate the wet season is widely used in subtropical daily water balance models (e.g., Kawal and Kassam 1977; Bunting *et al.* 1982). For the purpose of this study, ten days constitutes a period of sufficient length to eliminate many 'false starts'. The choice of a period of 15 days with $w_i = 0$ to define a break within the wet season is supported by meteorological definitions of drought. There are also local traditions which support this choice. For example, there is the custom in parts of central Sudan of dividing the year into 28 periods of 13–14 days, each of which is traditionally regarded as possessing a 'typical' weather type (Ibrahim 1984).

The choice of 0.5 for the crop coefficient can be justified from experimental work. Measurements of water use of a wide range of semi-arid crops has shown that the actual evapotranspiration (ET_p) at sowing which ensures good crop establishment is on average 0.48 ET (Kawal and Kassam 1977: 75). This figure of 0.5 is supported by both Benoit (1977) and Doorenbos and Kassam (1979) who show that important physiological changes are induced in many crops if water input fails to reach 0.5 of ET_p demand. Agnew (1982), in a detailed field study of an area in south-west Niger with a similar climatic and soil environment to west-central Sudan, derived crop coefficients of 0.11 for pre-season bare soils and 0.70 for mid-season rainfed crops. A single value of 0.5 for the duration of the growing season therefore appears acceptable.

Breaks within the wet season were classified according to whether they occurred in early, mid- or late season. These three stages of the wet season were defined as the periods in which one-third of the median annual rainfall (MAR) would on average occur for each respective station. These dates were determined on the basis of the average daily rainfalls calculated from the complete record at each station and are indicated in Table 16.1 as day numbers. This classification is valuable in distinguishing between the significance of breaks. There are two critical periods when water shortage can cause wilting of plants and reduce yields (Benoit 1977;

Table 16.1 Summary characteristics of the 12 long-term daily rainfall series

Station	Latitude	Length of record	Complete years	Median annual rainfall (mm)	Mean annual number of >1 mm falls	Mean date by which 1/3 and 2/3 of MAR falls	
						1/3	2/3
Khartoum	15°45′N	1900–86	85	141	15	203	220
El Geteina	14°52′N	1906–86	80	183	14	202	222
Abu Deleiq	15°55′N	1906–86	77	197	13	202	223
Kamlin	15°05′N	1906–86	79	222	18	202	225
El Fasher	13°37′N	1917–86	70	252	27	201	220
Bara	13°42′N	1908–86	78	266	20	200	226
Ed Dueim	13°59′N	1902–86	84	286	24	200	225
Manaqil	14°15′N	1905–86	76	304	20	199	222
El Obeid	13°10′N	1902–86	85	353	32	198	226
Kosti	13°10′N	1931–86	54	382	34	198	224
En Nahud	12°42′N	1912–86	74	387	33	196	227
Sennar	13°33′N	1925–86	59	439	37	193	220

NB. day 200 = 19 July
       day 225 = 13 August.

Bunting *et al.* 1982). The first occurs immediately after sowing when a dry spell can halt germination, necessitating resowing if this is possible. This period corresponds to early-season breaks which may be thought of as 'severe' false starts to the wet season ('minor' ones having been eliminated by the model definition). Again, local traditions support this idea with the period 191–203 (10–22 July) being regarded as the critical 'natural period' for determining the success or failure of crop yields (Ibrahim 1984). The second occurs in the middle of the growing season during flowering when a dry spell can seriously reduce yields (Bunting *et al.* 1982). This period corresponds to mid-season breaks.

Final calibration runs of the model using the above definitions produce one difficulty. For the wetter stations the daily water balance (w_i) was remaining positive well into January of the following year, long after the rains had actually stopped. This resulted in unrealistically long wet seasons. To discriminate against such locations, the initial assumption of an infinite soil moisture capacity was replaced by a nominal field capacity of 150 mm. This coincides well with estimates by Bunting *et al.* (1982) and detailed empirical water balance studies by Agnew (1982) in south-west Niger.

Data

Long-term series of daily rainfalls were obtained for 12 climatological stations in central Sudan between 12° and 16 °N (Figure 16.1). Data were extracted manually from unpublished sources in the Climatological Archive Office in Khartoum, part of the Sudan Meteorological Service. The data were extracted up to 1986, although two stations failed to report beyond 1983. The length of record for each station varied from 56–87 years and key rainfall characteristics are summarised for each station in Table 16.1. Potential evapotranspiration (ET_p) values were obtained from Awadulla (1983). He calculated mean monthly ET_p totals for nearly 200

Sudanese stations based on the period 1951–80, adapting Penman's formula to accord with Sudanese conditions. The model therefore fails to consider daily and yearly variations in ET_p, yet these are likely to be minimal in relation to variations in rainfall amounts. The wet season model defined above was applied to the 12 daily rainfall series. The onset, termination and duration of the wet season were determined for each year according to the model specifications and breaks within the wet seasons were identified.

Results

Figure 16.2 illustrates the variability of the onset and termination of the wet season for all 12 stations based on the complete record available for each. It should be noted that in the calculation of the variability of these dates, years in which a wet season failed to materialise (termed 'null starts' and with an effective value of zero) had to be excluded. This is only relevant for the more northerly stations. The length of the wet season steadily increases with median annual rainfall (MAR). There is no apparent change in the variability of onset date as rainfall increases, the inter-quartile range being consistently 25–35 days. The termination date shows as much variability as the onset date, if not more. For the drier stations, the interquartile range for the termination date is consistently higher than for the onset date.

Changes in the frequency of null starts and breaks

The question posed at the outset of this paper was whether there has been any systematic change in the structure of the wet season during the present century. First, the distribution of null starts is investigated and, second, the distribution of breaks within the wet season. Their distributions during the present century are presented in matrix form, with the stations listed in order of increasing MAR (Figure 16.3).

The distribution of null starts broadly follows the fluctuations in annual rainfall in central Sudan (as noted by Trilsbach and Hulme 1984). The lowest frequency of wet season failures occurred in the 1930s and 1950s and the greatest number in the 1910s, 1940s, 1970s and 1980s. The most recent dry period since the end of the 1960s, however, has clearly had the highest frequency of null starts rather than earlier drought episodes in the twentieth century. For the full twentieth-century record, the probability of a null start per station-year is 0.052. The probability of a certain period receiving any given number of null starts can be determined using the binomial distribution with $p = 0.052$. Eleven null starts occurred in the early drought period 1901–19 ($p = 0.11$) but 21 in the recent drought period 1968–86 with a probability of occurrence under the hypothesis of a random distribution of only 0.004. Of greater importance, however, is the fact that until the 1970s null starts had only occurred at stations with MAR less than 225 mm, while within the last 15 years null starts have been recorded for the first time at stations with MAR greater than 250 mm: at Bara in 1985, Ed Dueim in 1974 and 1984 and Manaqil in 1977.

In contrast, the frequency of breaks within the wet season reached a maximum in the wet 1940s and 1950s and a minimum in the dry 1970s (Figure 16.3b). This suggests that fewer breaks occur in times of drought. However, when the behaviour

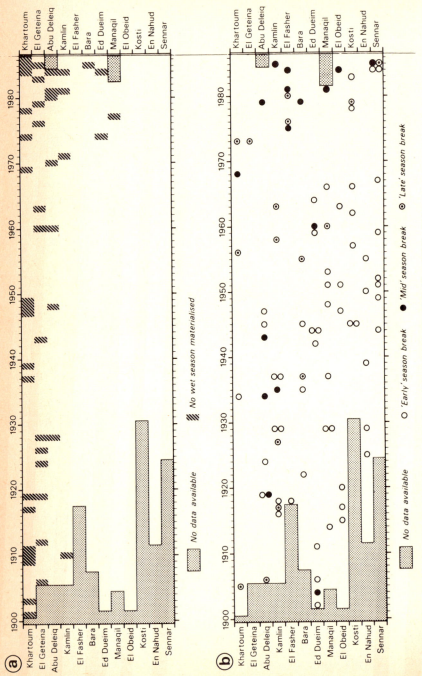

Figure 16.2 Variability in the onset and termination of the wet season for 12 rainfall stations

Note: based on data to 1983 only.

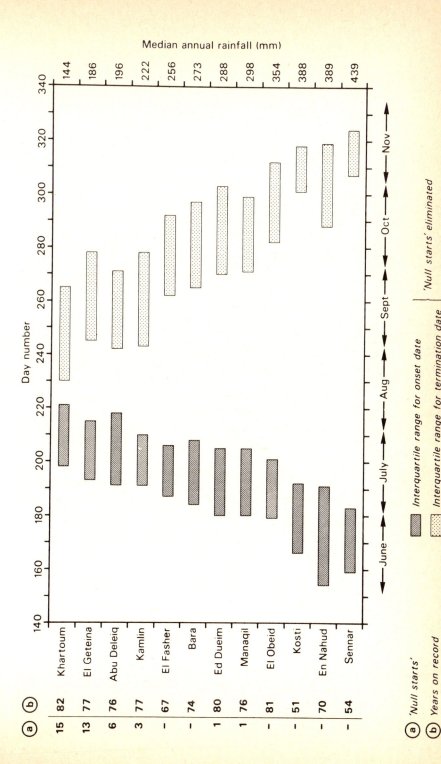

Figure 16.3 Annual wet season structure matrices for 12 rainfall stations, data to 1986: (a) null starts; (b) breaks within the wet season

Table 16.2   Frequencies of wet season breaks in 1901–67 and 1968–86, with probabilities of such frequencies under a binomial distribution

Period	Number of station-years	Frequency of breaks (probability)		
		Early season	Mid-season	Late season
1901–67	693	53 ($p = 0.07$)	6 ($p = 0.05$)	10 ($p = 0.37$)
1968–86	221	4 ($p = 0.002$)	10 ($p = 0.005$)	6 ($p = 0.18$)

of different categories of breaks is investigated contrasting patterns emerge (Table 16.2). Since 1968, only four early-season breaks have occurred, a situation with a probability of only 0.002 under a binomial test. Also since 1968, however, the frequency of mid- and late-season breaks has increased with ten and six occurrences in each respective category. The large number of mid-season breaks in 1968–86 is significantly different (at 99%) from that expected under a binomial distribution (Table 16.2). In the most recent dry period, therefore, false starts to the wet season have become less frequent (perhaps because of a later overall onset to the wet season), whilst the reliability of rainfalls in mid-season especially has decreased.

Changes in onset, termination and duration

In view of the apparent close association between annual rainfall and the wet season parameters, the series were investigated for trends during the present century. Has the recent potential shortening of the wet season resulted in a statistically significant trend? Mann–Kendall's rank correlation coefficient which tests for trend in a time series was calculated for each series. Two-tailed tests were used in each case (i.e. no a priori assumptions were made about whether shortening or lengthening had occurred). As with previous analyses null starts had to be excluded from the onset and termination series, but were included in the duration series.

The results of the Mann-Kendall tests are shown in Table 16.3 together with the results of the same test applied to the annual rainfall series for each station. Significant shortening (at 95%) of the wet season occurred at six stations, five of them in the zone of 200–300 mm MAR. This shortening was clearly associated with an earlier termination of the wet season rather than with a later onset date. El Fasher is the exception here, with a significant contracting at both ends of the wet season. At the driest (MAR < 200 mm) and wettest (MAR > 300 mm) stations few significant trends were recorded. It is important to note that of the six stations which recorded a significant falling trend in annual rainfall, five also recorded a significant shortening of the wet season.

Spatial pattern of wet season duration, 1983 and 1986

Expanding the network of stations from 12 to over 30 for two sample years, 1983 and 1986, enabled the spatial pattern of wet season duration to be considered. These two years both fall within the period of drought resurgence in the mid-1980s. The standardised annual rainfall z-score for central Sudan in 1983 was -1.24 (an extremely dry year and the second driest on record) and in 1986, -0.73 (a slightly

Table 16.3   Mann–Kendall's rank correlation statistic τ, to test for trend in three wet-season parameters and annual rainfall

Station	Median annual rainfall (mm)	Annual rainfall	Onset	Termination	Duration
Khartoum	141	−0.08	−0.03	−0.08	−0.07
El Geteina	183	−0.16*	0.05	−0.10	−0.12
Abu Deleiq	197	−0.06	0.02	0.02	−0.08
Kamlin	222	−0.24**	0.13	−0.16*	−0.25**
El Fasher	252	−0.29**	0.21**	−0.23**	−0.27**
Bara	266	−0.16*	0.02	−0.16*	−0.17*
Ed Dueim	286	−0.22*	0.12	−0.22**	−0.21**
Manaqil	304	−0.20**	0.10	−0.18*	−0.16*
El Obeid	353	−0.09	0.04	−0.07	−0.07
Kosti	382	−0.11	0.05	−0.10	−0.14
En Nahud	387	−0.12	0.13	−0.12	−0.19*
Sennar	439	−0.01	−0.01	0.02	0.03

* $p < 0.05$ ** $p < 0.01$ (two tailed).

less extreme year, but still typical of the post-1960s regime). Figure 16.4 presents the duration data for these two years and also the change in wet season duration between the two years. If one makes the assumption that the duration of the wet season is a better indicator of crop performance than annual rainfall total (justifiable from the above discussion specifying the wet season model), then interesting spatial variations in duration occur.

While duration is marginally longer (10–20 days) in 1986 than 1983 over east and north-central Sudan, in western and west-central Sudan (a main rainfed cultivation zone) the wet season was 35–50 days longer in 1986 (Figure 16.4c). Of even more interest, however, is the area in south-central Sudan just west of the White Nile and again a major rainfed crop zone. Here, despite the much better regional annual rainfall value (−0.73 compared with −1.24), the 1986 wet season was substantially *shorter* by 50–60 days than 1983. Very different regional signals in wet season performance are therefore indicated by this type of analysis, highlighting localised areas where crop performance may be improved or deterred.

This type of parameter (wet season duration) may also have more utility than annual rainfall totals in accounting for the distributions of annual primary production defined by remote vegetation monitoring methodologies using NOAA AVHRR data (e.g., Hielkema *et al.* 1986). Indeed, these authors point out the potential of monitoring *effective* rainfall by such techniques and wet season duration may well be the sort of rainfall parameter that is represented by such vegetation indices. There is scope here, then, for more extensive comparative work between wet season duration as defined above and annual primary production defined by satellite vegetation indices.

Discussion

This paper has investigated changes in wet season structure in central Sudan during the twentieth century, based on 12 long-term series of daily rainfalls. The results

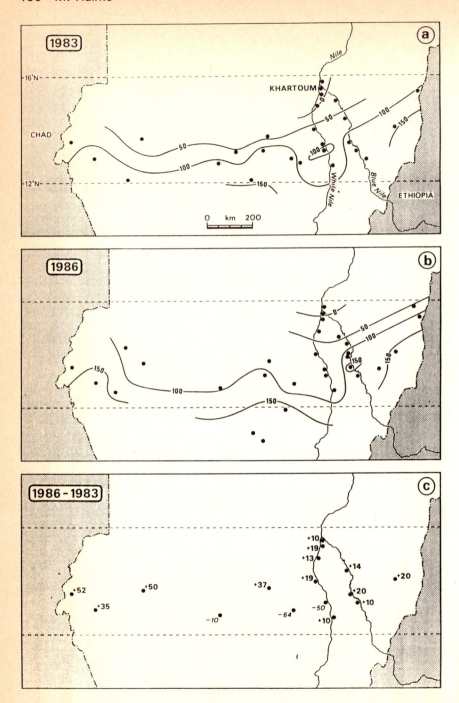

Figure 16.4 Duration of wet season in days (a) 1983; (b) 1986; (c) change in duration, 1986 minus 1983

presented above suggest that annual rainfall fluctuations are expressed consistently in fluctuations in the timing of the wet season as revealed by trend analysis of the time series (Table 16.3). Stations which displayed a significant falling trend in annual rainfall have also seen a significant shortening of the wet season during the century. This shortening is due more to early termination rather than later onset. Conversely, no significant shortening of the wet season has occurred where no trend exists in annual rainfall.

This covariance of annual rainfall with wet season duration rather contradicts the suggestion of Nicholson and Chervin (1983) and Dennett *et al.* (1985) that drought in the west African Sahel has not been associated with a simple shortening of the wet season. Whilst the increased frequency of mid- and late-season breaks in recent dry years (Figure 16.3) supports their claim of marked reductions in peak-system (August) rainfall, additional evidence is supplied here of systematic changes in the onset and termination of the wet season.

The drought years since the mid-1960s have displayed somewhat different wet season characteristics to the earlier droughts of the 1910s and 1940s. The contraction of the wet season has been more severe in recent years and this has resulted in three stations experiencing no wet season for the first time this century: Ed Dueim in 1974 and 1984, Manaqil in 1977 and Bara in 1985. The MAR of these stations is 250–300 mm, which puts them inside the recommended 'agronomic dry boundary'. The penetration of null starts into these agriculturally marginal latitudes represents a serious deterioration of growing conditions in central Sudan.

For marginal regions such as central Sudan, Parry and Carter (1984) outlined the value of focusing on changes in the frequency of extreme events rather than simply on long-term changes in mean climate. By using daily rainfall data and by investigating climatic change in terms of wet season structure (timing, null starts and breaks), this chapter has demonstrated the value of such an approach in understanding climatic impact in the Sahel. This analysis in central Sudan has revealed the anomalous behaviour of the critical 200–300 mm zone, the greater sensitivity of wet season termination date than onset date to annual rainfall changes and the decreased reliability on the internal structure of the wet season in the recent drought period. It has also contended that wet season duration may be a suitable parameter to relate to vegetation indices constructed from remotely sensed data and further comparative work here is intended.

Acknowledgements

The author thanks the staff of the Archive Office of the Sudanese Meteorological Service, in particular Taj el Deen, for assistance with the acquisition of data; Dr S.A. Awadulla for the use of potential evapotranspiration data; Dr R.P.D. Walsh and Dr M.D. Dennett for helpful comments on earlier drafts; and Gustav Dobryznski for the preparation of the illustrations.

References

Agnew, C.T. 1982. 'Water availability and the development of rainfed agriculture in southwest Niger, West Africa', *Trans. Inst. Brit. Geogr.* 7, 419–57.

Awadulla, S.A. 1983. *Potential evapotranspiration over the Sudan*, Scientific Note No. 3,

Sudan Meteorological Service, Khartoum.

Benoit, P. 1977. 'The start of the growing season in northern Nigeria', *Agric. Met.* 18, 91–9.

Bunting, A.H., Dennett, M.D., Elston, J., Milford, J.R. 1976. 'Rainfall trends in the West African Sahel', *Q.J.R. Met. Soc.* 102, 59–64.

Bunting, A.H., Dennett, M.D., Filbee, B., Rodgers, J.A. 1982. 'Water relations of groundnut crops' paper presented at The international symposium on groundnuts, June, International Groundnut Council, Banjul, Gambia.

Dennett, M.D., Elston, J., Rodgers, J.A. 1985. 'A reappraisal of rainfall trends in the Sahel', *J. Climatol.* 5, 353–61.

Doorenbos, J., Kassam, A.H. 1979. *Yield response to water*, FAO Irrigation and drainage Paper no. 33, FAO, Rome.

Gregory, S. 1983. 'A note on mean seasonal rainfall in the Sahel, 1931–60 and 1961–80', *Geogr.* 68, 31–6.

Hammer, R.M. 1968. 'A note on rainfall in Sudan', *Weather* 23, 211.

Hielkema, J.U., Prince, S.D., Astle, W.L. 1986. 'Rainfall and vegetation monitoring in the savanna zone of the Democratic Republic of Sudan using the NOAA Advanced Very High Resolution Radiometer', *Int. J. of Remote Sensing* 7, 1499–1513.

Hulme, M. 1986. 'The adaptability of a rural water system to extreme rainfall anomalies in central Sudan', *Appl. Geogr.* 6, 87–103.

Ibrahim, F. 1984. *Ecological imbalance in the Republic of the Sudan*, Bayreuth, Druckhaus Bayreuth Verlagsgesellschaft mbH.

Kawal, J.M., Kassam, A.H. 1977. *Agricultural Ecology of Savanna*, Oxford, Clarendon Press.

Lamb, P.J. 1983. 'Sub-Saharan rainfall update for 1982: continued drought', *J. Climatol.* 3, 419–22.

Nicholson, S.E. 1979. Revised rainfall series for the West African subtropics', *Mon. Wea. Rev.* 107, 620–3.

Nicholson, S.E., Chervin, R.M. 1983. 'Recent rainfall fluctuations in Africa — interhemispheric teleconnections', in A. Street-Perrott and M. Beran (eds) *Variations in the Global Water Budget*, Dordrecht, D. Reidel.

Parry, M.L., Carter, T.R. (eds) 1984. *Assessing the Impact of Climatic Change in Cold Regions*, Summary Report SR-84-1, International Institute for Applied Systems Analysis, Laxenburg.

Pedgley, D.E. 1969. 'Diurnal variation of monsoon rainfall', *Met. Mag.* 98, 97–107 and 129–34.

Trilsbach, A., Hulme, M. 1984. 'Recent rainfall changes in central Sudan and their physical and human implications', *Trans. Inst. Brit. Geogr.* 9, 280–98.

Chapter 17

Rainfall data bases and seasonal forecasting in eastern Africa

G. Farmer   University of East Anglia, United Kingdom

Introduction

This chapter is concerned with the collection and analysis of African rainfall data. Perhaps the first question should be, why are we focusing on Africa? The past 40 years have shown some of the possible vagaries of climate, both in the long-term changes of the Sahelian region (e.g., Lamb 1983; Nicholson 1985), and in the short-term catastrophic impact of a failure of seasonal rainfall (Glantz 1987). Given that most national economies in tropical Africa are agriculturally based, the importance of climate and particularly rainfall is easy to appreciate. The roles of climate, climatic variability and climatic change then become worthy of consideration from more than a scientific viewpoint.

Climatological analyses need data and quantitative rainfall measurements have been taken over the past five centuries, perhaps as early as the 1440s in Korea. For Africa we can use rainfall series that were begun in the late nineteenth century. In practice many of the early analyses of African rainfall were restricted to relatively short data series for a few locations. In recent times the movement has been towards utilising large collections of data, in a data base. There is a need, therefore, to ensure that the rainfall data base, which is more often used by researchers who are divorced from the original data collection, is accurate, error-free and as up to date as possible.

Most analyses of large amounts of African rainfall data have been undertaken where computer technology is readily available, generally in Europe, Japan and the USA. In recent years, however, the trend has been to install new computers with increased power in African meteorological agencies and similar institutions. One aim of this chapter is to highlight some of these developments and to show their potential for analysis and research. The emphasis here will be placed on eastern and southern Africa, based on a proposed research project, but the implications drawn are applicable at the continental scale.

Until a data base is analysed it is in fact of minimal use. Examples are given below of data-base development and subsequent analysis. These are concerned with two broad approaches. The first is to use the data base to examine spatial and temporal variations in annual rainfall amounts. The other use presented here is to develop regression models for forecasting seasonal rainfall totals, with varying degrees of success.

Rainfall data bases

Rainfall data are, on the whole, collected by national meteorological agencies. An individual observer at a recording site measures the rainfall over a set period of time, for example a day or a month, and records it in a book. The data are periodically sent to the national agency headquarters for incorporation into a national data base and storage.

Additional data bases may be developed by individual research workers, perhaps specialising in a particular geographical region. There also exist, outside Africa, several global and continental-scale rainfall data bases. The starting point for some of this research has been the combined data bases of the National Center for Atmospheric Research (NCAR) at Boulder, Colorado, USA (Jenne 1975), and the extensive African rainfall data base of S.E. Nicholson (SEN), now at Florida State University, Tallahassee, USA (e.g., Nicholson 1980). NCAR collates monthly rainfall data on a global scale that are later published in *Monthly Climatic Data for the World*. The NCAR and SEN databases have been merged at NCAR giving, for Africa, a coverage of over 1,300 sites.

In any data base of any size there will be errors. The NCAR–SEN data base is no exception, but neither is it particularly worse than other data bases. Potential errors can be seen throughout the 'life' of an individual data value. For example, at the recording site there are three sources of error. The gauge may be giving an incorrect catch, for one or more reasons, the observer may misread the amount, and the amount may be written incorrectly in the log. Thereafter, every time a number is copied from one medium to another an error may occur.

A common error relates to units of measurement. Rainfall totals are often reported in tenths of a millimetre, but the NCAR Monthly Climatic Data for the World database is in whole millimetres. Order of magnitude errors can occur, and have been found, where somewhere in the transition from original readings to final data base, a value such as 86.1 mm becomes 861 mm rather than being rounded to 86 mm. Other data-input problems include transposing of numbers, duplication of records and omission of data because they did not reach part of the recording system by the required date. Although the data base reports 'missing data' the data are, quite often, in existence in the country of origin. Related to this topic, there are important concerns about economic climates and the run-down of recording networks and closure of stations. This is not addressed here, but the problem is of relevance to many African countries.

Another important area is that of data homogeneity. That is to say that any part of a record for a location should be comparable with any other part. This is not always the case. For example, site changes in the NCAR data set are not documented, with records for two different sites within one city being simply joined together. This gives the immediate potential of a source of non-climatic changes in rainfall totals or characteristics. Other non-climatic problems may rise from the deterioration of a site in relation to the exposure of the instrument or recording procedures.

The above discussion gives a brief outline of some of the difficulties involved in building any data base (see Farmer and Wigley 1985, for more details). A further aspect in relation to African rainfall is that rainfall data bases have been compiled, often by necessity for computer facilities, away from Africa. This is a lamentable situation as it has deprived the national agencies of the broader data availability and subsequent analysis. Recent developments, however, are addressing this situation.

With funding from various parts of the United Nations Development Programme two drought monitoring centres (DMC) are being established, one in Nairobi, Kenya and one in Harare, Zimbabwe. The Nairobi DMC is already in operation. At the continental scale, the UN Economic Commission for Africa has recently designated Niamey, Niger, as the site for the proposed African Centre for Meteorological Applications and Development (ACMAD). There will no doubt be contact between ACMAD and AGRHYMET, the monitoring, research and training facility established by the CILSS Sahelian countries after the severe drought of 1972 (also based in Niamey). There is also the World Meteorological Organisation's CLICOM project which has the aim of facilitating the collection and analysis of data by standardising national agencies with IBM microcomputers.

One of the roles of the DMCs will be to establish reliable, homogeneous data bases. These will initially be for monthly rainfall and will concentrate, at Nairobi and Harare, on eastern and southern Africa. Monthly rainfall totals provide a compromise between coarseness of scale and speed of input. Further development of the data bases would increase the time resolution for rainfall with shorter-period data, e.g., daily or pentad (important for agricultural studies), incorporate other climatic parameters and expand the station distribution. Further discussion of these aspects can be found in Davy (1982) and WMO (1986). If funding becomes available it is hoped that the Climatic Research Unit of the University of East Anglia, Norwich, UK, will be involved in collaborative work on the building and analysis of the rainfall data bases.

There are clear advantages in developing an African rainfall data base in Africa. The product can be geared to the national and regional requirements with regard to station network. A multinational institute could ensure that late reports from countries within the region are not simply lost as missing data. Moreover, and possibly most important of all, there is a greater possibility for research to be carried out by Africans in Africa. This is not to ignore the beneficial aspects of collaboration with scientists from outside the region. It gives, however, a greater opportunity for local expertise to be developed and extended.

Data base application: analysis

Once a reliable data base has been prepared, it is to some people a thing of beauty. However, until the data are applied to a problem they are of limited value. Example analyses given here were chosen for their contemporary interest, and to show possible national social and economic benefits from such studies.

Spatial and temporal variability of rainfall

A recent analysis has used the NCAR–SEN data base to construct time series of rainfall indices for different regions and zones of the northern hemisphere (Bradley et al. 1987a; Jones, Chapter 3, this volume). One region presented by Bradley et al., northern Africa and the Middle East (NA/ME), covers the region 15–35 °N, 20–65 °W. Their method was to transform all the station data to probability units based on the gamma distribution. Transformed in this way the data are much more readily gridded onto a regular grid network. The chapter by Jones gives more detail.

The method used here calculates mean normalised anomalies (MNA) for a geographical region. This is performed by normalising each station with respect to its long-term mean and standard deviation; 1931–60 is used here as the base period. The MNA for any particular year is then the average normalised value of those

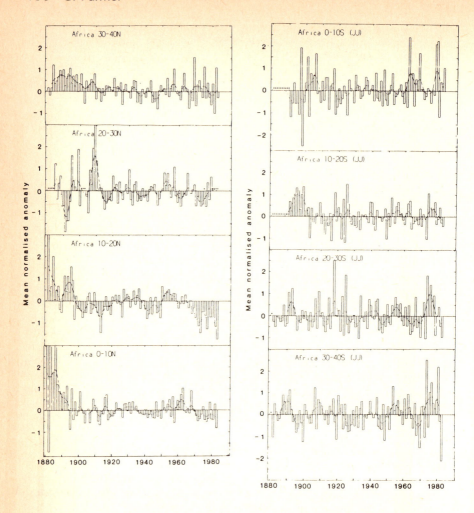

Figure 17.1   Time series graphs of annual rainfall mean normalised anomalies for 10°
latitude bands over Africa, 1881–1983

Note: For description of method, see text. The smooth lines are nine-term padded binomial filters.
The number of stations available for analysis varies year by year. Asterisks indicate years with no
stations. Southern hemisphere values are calculated from July to June and plotted in the year of the
January. Values exceeding +3 standard deviations are plotted *at* +3 σ.

stations that have annual data. A discussion of the method is given in Katz and
Glantz (1986). For the purpose of this chapter the results of this procedure and that
of Bradley *et al.* (1987a) are broadly comparable.

If we construct time series for 10° latitude bands in the NCAR–SEN data base
(Figure 17.1), we see that the 'north African' decline is dominated by the Sahelian
latitudes of 10–20 °N (see also Lamb 1985; Nicholson 1985: Farmer and Wigley
1985). The decline is less marked north of 20 °N, although for 20–30 °N no
stations in the NCAR–SEN data base gave complete annual data for the period

1980–3. In the southern hemisphere none of the four series shows periods of prolonged surplus or deficit. The southern hemisphere annual totals are computed from July to June and dated to the year of the January. It is important to note that annual totals for some regions near the equator will be the total of more than one rainy season. A continental-scale data base can be used, therefore, to provide useful large-scale measures of rainfall activity. Smaller-scale data bases for selected regions, generally with greater station density, have similarly been used to look at the rainy season in more detail (Dennett *et al.* 1985; Hutchinson 1985; Eldredge *et al.* 1988).

Forecasting of seasonal rainfall totals

As stated previously, most of the national economies in tropical Africa are agriculturally based. There are clear economic benefits, therefore, in a reliable forecasting system for the amount of rainfall in a growing season. We are considering here the time-scale of the order of a few months. Meteorological forecasting of a few days is also of benefit to agriculture, but is not the topic of interest here.

There has been much interest recently in teleconnections between El Niño/Southern Oscillation (ENSO) events and rainfall and temperature anomalies in different parts of the world. For a discussion of ENSO, see Rasmussen and Carpenter (1982) and Caviedes (Chapter 22, this volume). Summaries of global-scale teleconnections can be found in Rasmussen (1984), Nicholls (1987) and Ropelewski and Halpert (1987). African ENSO-rainfall studies include Ogallo (1987) and Nicholson and Entekhabi (1986). Figure 17.2 shows rainfall indices for particular seasons (details given in captions) for three regions in Africa: Figure 17.2a, the eastern equatorial (EEQ); Figure 17.2b, south-eastern (SEA) regions of Ropelewski and Halpert (1987); and Figure 17.2c, the Kenya coast of Farmer (1988). Solid bars in all three series show ENSO warm event years. In Figure 17.2c the striped bars show ENSO cold event years, from Bradley *et al.* (1987b). A location map for the regions is given in Figure 17.3.

On the gross scale, Figure 17.2 suggests that ENSO events can be used as predictors for the three regions. For the Kenya coast short rains (September to November) the relationship has been investigated further. Full details are given in Farmer (1988); some of the more promising results are summarised here.

For the period 1943–84 regression equations were computed between seasonal values of a Tahiti–Darwin Southern Oscillation Index (SOI) (Ropelewski and Jones 1987) and the Kenya coast rainfall index for September to November. After removal of two outlier values in the rainfall series the correlation between June to August (JJA) SOI and September to November (SON) rainfall was −0.65. This relationship was tight throughout the range of SOI and rainfall values, and gives a 'predictive' equation that explains almost half of the variance (R^2 adjusted = 0.40) in SON rainfall by using the JJA Southern Oscillation Index. This has the beginnings of a useful tool. However, the relationship is weaker over the 1901–42 period, although this may be a real change resulting from a change in underlying mechanisms. Such aspects are being investigated further.

As an extension to this work a similar approach was taken using seasonal SOI series against rainfall series derived for the smaller areas in EEQ and SEA (Figure 17.3). In general terms the results were disappointing. Southern Uganda showed better results for the October to December season rather than that of March to May.

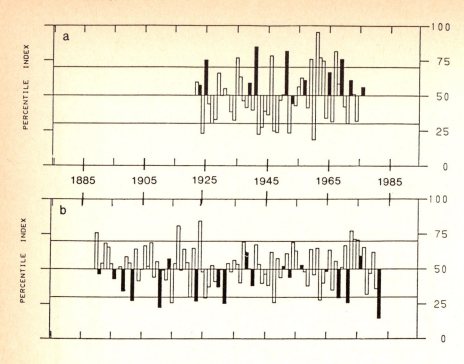

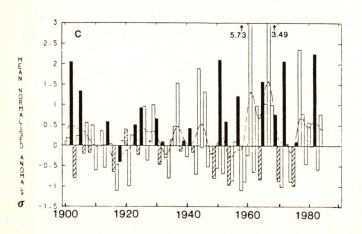

Figure 17.2   Seasonal rainfall series for different regions of Africa

Note: Figures 17.2a and 17.2b are from Ropelewski and Halpert (1987). These show percentile index values of (a) October to April rainfall in eastern equatorial Africa, and (b) November to May rainfall in south-eastern Africa. Figure 17.2c (from Farmer 1988) gives mean normalised anomalies for September to November on the Kenyan coast. All locations are shown in Figure 17.3. Solid bars indicate ENSO years. In (c) the striped bars indicate ENSO cold years (from Bradley *et al.* 1987b). Coarse relationships between ENSO and seasonal rainfall are readily apparent.

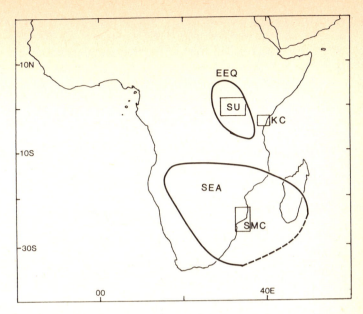

Figure 17.3 Location map for series in Figure 17.2 and additional analyses

Note: While in Kenya coast (KC) shows a predictive relationship between a seasonal Southern Oscillation Index and September to November rainfall, results in southern Uganda (SU) and the southern Mozambique coast (SM) were not as promising. The latter results are in contrast to the series for eastern equatorial (EEQ) and south-eastern Africa (SEA) presented in Figure 17.2.

In these, and the southern Mozambique coastal region, there was markedly little of the rainfall variance being explained by seasonal SOI, a somewhat surprising result given the relationships in Figure 17.2. These, and other results, are being subjected to further analysis. It would appear that while the large-scale series reflect gross movements of ENSO, the smaller-scale sub-units chosen do not exhibit such detail. (A similar discussion for India is given by Gregory; Chapter 20, this volume).

Summary

A reliable, homogeneous, extensive and up-to-date data base is of essential importance in climatic research. While Africa is better served than some other parts of the world there is still room for improvement. Part of this improvement can now be undertaken in new and planned African research and monitoring centres such as the drought monitoring centres in Nairobi and Harare, and the African Centre for Meteorological Applications and Development in Niamey. Through national and international collaborative work it is hoped that new impetus may be given to the study of rainfall and drought in Africa.

Data bases are, however, not static. They should be constantly evolving, increasing the number of stations and updating those in the data base. Most important of all, they should be the subject of analysis. Some introductory topical analyses have been presented that show the value of an African rainfall data base in monitoring spatial and temporal patterns of change. Marked contrasts can be seen between long-term fluctuations in Sahelian latitudes and those to the north and south.

Seasonal forecasting of rainfall has been investigated using simple regression techniques with a Southern Oscillation Index as an indicator of ENSO activity. While the method gives promising results for coastal Kenya, a similar approach in Uganda and Mozambique has been found to be less fruitful. Given that a coarse ENSO signal seems to be present in eastern equatorial and south-eastern Africa (Figures 17.2 and 17.3) it is an interesting reflection that moving to the finer detail for coarse relationships does not necessarily mean that relationships will be as strong. While perhaps this is a realistic comment, it makes the goal of useful seasonal forecasting more difficult to achieve. It is hoped that a combination of the methods described here and those used in Brazil (Ward *et al.*, Chapter 21, this volume) will provide insight into this problem in future research.

Acknowledgements

Supplementary to the NCAR–SEN data base used here, additional rainfall data were supplied by the National Meteorological Departments of Kenya and Uganda, to whom I am extremely grateful. Thanks are also due to P.D. Jones for useful comments on this chapter, and to the UK Overseas Development Administration which provided financial support for part of this work.

References

Bradley, R.S., Diaz, H.F., Eischeid, J.K., Jones, P.D., Kelly, P.M., Goodess, C.M. 1987a. 'Precipitation fluctuations over Northern Hemisphere land areas since the mid-19th century', *Science* 237, 171–5.

Bradley, R.S., Diaz, H.F., Kiladis, G.N., Eischeid, J.K. 1987b. 'ENSO signals in continental temperature and precipitation records'. *Nature* 327, 497–501.

Davy, E.G. 1982. 'Summary of the climatological data situation in Africa'. In *Proceedings of the Technical Conference on Climate — Africa*, WMO no. 596, Geneva, World Meteorological Organisation, 177–87.

Dennett, M.D., Elston, J., Rodgers, J.A. 1985. 'A reappraisal of rainfall trends in the Sahel', *J. Climatol.* 5, 353–62.

Eldredge, E., Khalil, S. El S., Salter, C., Nicholls, N., Abdalla, A.A. Rydjeski, D. 1988. 'Changing rainfall patterns in western Sudan', *J. Climatol.*, 8, 45–53.

Farmer, G. 1988. 'Seasonal forecasting of the Kenya Coast Short Rains, 1901–84', *J. Climatol.*, in press.

Farmer, G., Wigley, T.M.L. 1985. *Climatic Trends for Tropical Africa, Research Report for the Overseas Development Administration*, Norwich, Climatic Research Unit, University of East Anglia.

Glantz, M.H. (ed.) 1987. *Drought and Hunger in Africa: Denying Famine a Future*, Cambridge, Cambridge University Press.

Hutchinson, P. 1985. 'Rainfall analysis of the Sahelian drought in the Gambia, *J. Climatol.* 5, 665–72.

Jenne, R. 1975. 'Data sets for climatological research', *NCAR Technical Report 1A-111*. Boulder, CO, National Center for Atmospheric Research.

Katz, R.W., Glantz, M.H. 1986. 'Anatomy of a rainfall index', *Mon. Wea. Rev.* 14, 764–71.

Lamb, P.J. 1983. 'Sub-Saharan rainfall update for 1982: continued drought', *J. Climatol.* 3, 419–22.

Lamb, P.J. 1985. 'Rainfall in sub-Saharan West Africa during 1941–83', *Zeitschrift für Gletscherkunde und Glazialgeologie* 21, 131–9.

Nicholls, N. 1987. 'The El Niño/Southern Oscillation phenomenon' in M. Glantz, R. Katz and M. Krenz (eds), *The Societal Impacts Associated with the 1982–83 Worldwide Climate Anomalies*, Boulder, CO, National Center for Atmospheric Research, 2–10.

Nicholson, S.E. 1980. 'The nature of rainfall fluctuations in subtropical West Africa', *Mon. Wea. Rev.* 108, 473–87.

Nicholson, S.E. 1985. 'Sub-Saharan rainfall 1981–84', *J. Clim. Appl. Met.* 24, 1388–91.

Nicholson, S.E., Entekhabi, D. 1986. 'The quasi-periodic behaviour of rainfall variability in Africa and its relationship to the Southern Oscillation', *Arch. Met. Geoph. Biocl.*, ser. A, 34, 311–48.

Ogallo, L.J. 1987, 'Teleconnections between rainfall in eastern Africa and global parameters'. In *Proceedings of the First Technical Conference on Meteorological Research in Eastern and Southern Africa*, Nairobi, Kenya Meteorological Department.

Rasmussen, E.M. 1984. 'El Niño: the ocean/atmosphere connection', *Oceanus* 27, 5–12.

Rasmussen, E.M., Carpenter, T.H. 1982. 'Variations in tropical sea surface temperature and surface wind fields associated with the souther Oscillation/El Niño', *Mon. Wea. Rev.* 110, 354–84.

Ropelewski, C.F., Halpert, M.S. 1987. 'Global and regional scale precipitation and temperature patterns associated with the El Niño/Southern Oscillation', *Mon. Wea. Rev.*

Ropelewski, C.F., Jones, P.D. 1987. 'An extension of the Tahiti–Darwin Southern Oscillation Index', *Mon. Wea. Rev.*

WMO (World Meteorological Organisation). 1986. 'Guidelines on the structure, management and operation of climate data centres', *WCP-99*, Geneva, World Meteorological Organisation.

Chapter 18

Synoptic circulation types and climatic variation over southern Africa

P.D. Tyson   University of Witwatersrand, Johannesburg, South Africa

Notwithstanding the important contribution of cumulus convective rainfall to total annual rainfall over southern Africa, the occurrence of rainfall on a daily scale is highly dependent on synoptic circulation changes. On longer time-scales, from several days to several years, the modulation of extended wet and dry spells by the circulation has been extensively investigated. A recent review is available (Tyson 1986). However, little effort, *per se*, has been directed to understanding the similarity in the synoptic forcing of daily rainfall and the forcing of wet spells on a near-decadal scale. Attention will be directed to this topic in this chapter.

Circulation classification

The first synoptic classification of southern African weather systems was that of Vowinckel (1955; 1956). Using a statistical approach similar to that advocated by Lund (1963) and employing daily synoptic charts, Longley (1976) classified weather types producing similar temperature and rainfall distributions. Erasmus (1980) used surface and 850 hPa synoptic conditions and incorporated satellite imagery into a three-way classification of tropical disturbances, temperate disturbances and convection-suppressing anticyclonic activity. Other classifications have been developed, including that of McGee (1974) for determining water vapour transport with different air-flow types; that of Garstang *et al.* (1981) for distinguishing between hail-producing and hail-suppressing conditions in the Transvaal Lowveld and that of Diab and Garstang (1985), for estimating conditions favourable for the generation of wind energy. The last two classifications are based on surface synoptic conditions, whereas McGee incorporated the vertically-integrated vapour transport and net wind direction between the surface and 300 hPa.

An alternative approach will be considered here. Using generalised circulation types and their attendant upper-level counterparts it is possible to distinguish three categories of circulation pattern: first, fine-weather and mildly disturbed conditions; second, tropical disturbances associated with tropical air flow, and third, temperate, mid-latitude disturbances associated with westerly air flow (Figure 18.1). Within the three categories ten circulation types have been recognised. Apart from the fine-weather types, two represent tropical forcing and six temperate forcing of rainfall. When the tropical and temperate circulations become coupled major cloud-band and rain swathes linking the tropics and middle latitudes are formed. The significance of these linking cloud bands has been demonstrated by Harrison (1984; 1986). Using satellite imagery and cloud occurrence as an indication of vertically integrated processes Harrison's generalised classification of summer

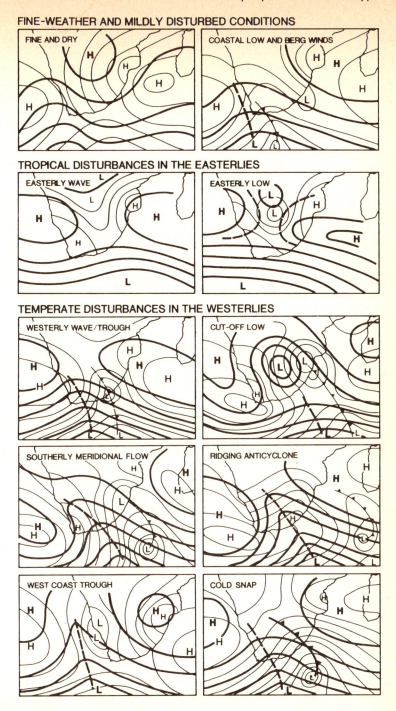

Figure 18.1 A schematic classification of southern Africa weather types based on circulation patterns at the surface (light lines) and 500 hPa (heavy lines)

SYSTEMS LINKED TO TROPICAL CIRCULATIONS

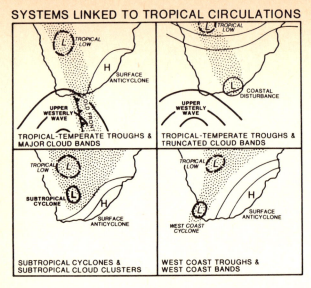

SYSTEMS WITHIN THE WESTERLY CIRCULATION

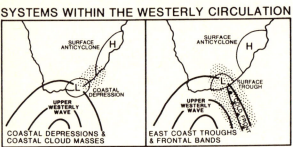

Figure 18.2 Generalised classification of circulation types and their attendant cloud systems

Note: Lower-level circulation is shown by light lines; upper level by heavy lines. Cloud cover is shown by stippling
Source: Harrison (1986).

rain-bearing synoptic systems for South Africa relates particular patterns of cloud organisation to surface and upper tropospheric circulation and to rainfall patterns determined by principal component analysis (Figure 18.2). The monthly frequencies of the various cloud systems that are indicative of the different rain-bearing systems are distinctively different (Figure 18.3). Over the central interior of southern Africa it is clear that major cloud bands linking the tropical and temperate latitudes are the principal contributors to rainfall and that they and other tropically induced systems vary with a clear annual cycle having a peak in summer. Temperate disturbances, on the other hand, contribute less by way of rain in the interior and peak in the frequency of their occurrence in spring and autumn in response to a semi-annual cycle. The annual forcing increases toward the tropics; the semi-annual dominates in middle latitudes.

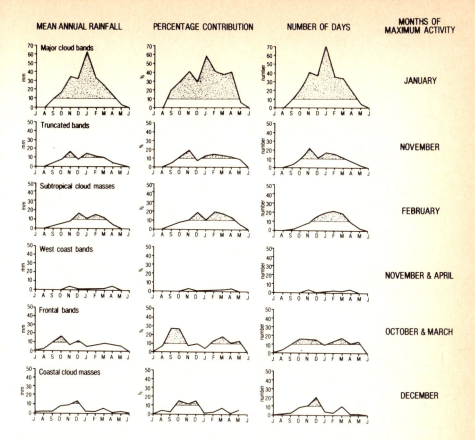

Figure 18.3   The annual march of significant rainfall resulting from circulation types producing the given cloud systems over the eastern Orange Free State

Note: Stippling is for effect only and indicate values exceeding 10 mm, 10 per cent or a frequency of 10

Source: Modified after Harrison (1984).

The importance of the upper-level circulation

It has long been known that the nature of the upper-air circulation is a major determinant of surface weather. Taljaard (1958) pointed out how differential advection between the surface and 500 hPa surface was largely responsible for the generation of line storms. Many other workers have shown the importance of the upper air flow in explaining particular extreme events. Recently Steyn (1984) demonstrated how the curvature of streamlines at 300 hPa over central South Africa determined the occurrence of isolated, scattered or general rainfall. A particular example of the controlling effect of upper air flow on the occurrence of rainfall is given in Figure 18.4. On both days analysed a similar surface circulation prevailed with an easterly wave appearing likely to be conducive to precipitation. On both days the thermal stability was similar: conditional instability occurred throughout the

Figure 18.4 Near-identical easterly wave surface circulation patterns on a rain and no rain day (area of rain shaded on 850 hPa chart) when the amount of conditional instability present was approximately similar

Note: On the rain day 500 hPa divergence in the wave at that level promoted uplift to realise the instability; on the no-rain day 500 hPa convergence in a ridge suppressed vertical motion and the occurrence of rainfall. Dry bulb and dew point temperatures are given on the tephigrams.

troposphere and potential instability was observed to the 600 hPa level. The vertical moisture distribution was likewise similar on the two occasions. On the rain day upper-level divergence prevailed over South Africa and together with low-level convergence to the east of the easterly trough produced the necessary uplift. By contrast, on the no-rain day a 500 hPa ridge of high pressure was associated with a convergent windfield at that level and prevented the development of any uplift.

Longer-duration rainy spells

In an analysis of 20 wet and 20 dry summer days, Triegaardt and Kits (1963) showed that if pressures at 500 hPa fell over the west coast (owing to the frequent occurrence of deep west coast troughs), then widespread rains fell over central South Africa (Figure 18.5). The corresponding dry spells showed the anomalous occurrence of anticyclonic conditions at 500 hPa. In a similar fashion Miron and Tyson (1984) have demonstrated the tendency for 500 hPa pressures to drop over the interior and to rise over the Gough Island region during wet months, wet seasons and even during extended spells of wet years (Figure 18.6).

It is possible to isolate particular wet months in which some of the specific synoptic types illustrated in Figure 18.1 occurred sufficiently frequently to impose their distinctive imprints on 500 hPa pressure anomaly patterns (Figure 18.7).

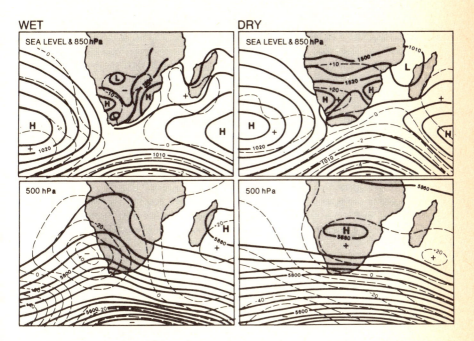

Figure 18.5 Mean sea level pressures (hPa), contours of the 850 hPa and 500 hPa surfaces (gpm) (solid lines) and departures from mean conditions (light broken lines) for 20 wet days in December 1960 and January 1961 and 20 dry days in January and February 1961

Source: Triegaardt and Kits (1963).

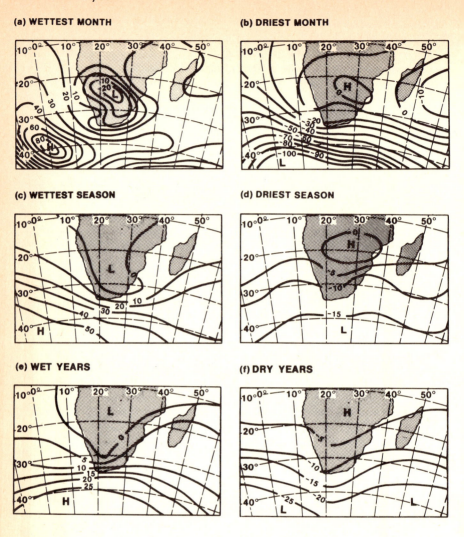

(a) WETTEST MONTH

(b) DRIEST MONTH

(c) WETTEST SEASON

(d) DRIEST SEASON

(e) WET YEARS

(f) DRY YEARS

Figure 18.6   Mean 500 hPa deviations (gpm) for: (a) the wettest January (1978); (b) the driest January (1969); (c) the wettest summer (1975/6); (d) the driest summer (1965/6); (e) all the wet months in the extended wet spell of 1972/73 to 1978/9; (f) all the dry months in the extended dry spell of 1963/4 to 1970/1

Note: The symbols H and L indicate relative states only.

The degree to which the actual curvature of streamlines in the upper troposphere affects precipitation over the southern Africa region may be quite dramatic. As Steyn (1984) shows, as the mean flow becomes increasingly cyclonic so above normal rainfall increases over the central plateau (at the same time the frequency of occurrences of the systems producing the enhanced precipitation decreases) (Figure 18.8). In each case the predominant forcing of the rainfall is the upper-level divergence occurring over the area. By contrast, below normal rainfall is associated

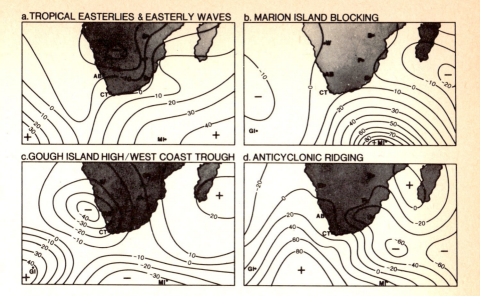

Figure 18.7 Examples of 500 hPa anomaly fields (departures from normal, gpm) associated with wet periods extending from five to thirty days (a) for January 1974 to illustrate the effect of easterly waves and tropic easterlies, (b) for February 1970 to illustrate Marion Island blocking, (c) for four-day wet periods in 1961 to illustrate Atlantic high forcing and the Gough Island high/west coast trough and (d) for 7–12 March 1963 to illustrate the effect of anticyclonic ridging south of the Cape

Sources:
(a) Taljaard (1981).
(b) Taljaard (1981).
(c) Taljaard and Kits (1963).
(d) Gouws (1963) and Tyson (1984).

with convergence in the upper air. As the degree of convergence behind the upper trough increases, so rainfall diminishes. It is clear that the wave configuration of the upper-level air flow is crucial in modulating rainfall. Any circulation adjustment leading to the frequent occurrence of upper-level troughs over the west coast will be conducive to enhanced rainfall over southern Africa. Any adjustment displacing the locus of most frequent trough occurrence to a position to the east of Natal will lead to the general diminution of rainfall over the region.

Near-decadal wet and dry spells

To date it has proved logistically impossible to extend the above types of analysis to several decades. Instead *annual* geostrophic flow indices have been developed for pairs of stations over the subcontinent and between the subcontinent and Gough and Marion Islands and Mauritius for the 20-year period 1958–1977. These indices give measures of the synoptic types occurring on a daily scale. For instance, the Gough Island to Cape Town (or Alexander Bay) index gives a measure of a west

above normal rainfall

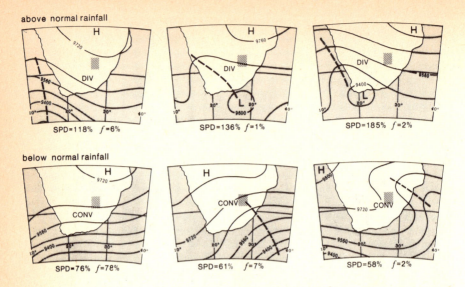

Figure 18.8   Upper level, 300 hPa, circulation patterns associated with specific precipitation densities, SPD,* occurring with above- and below-normal rainfall over the eastern Orange Free State (shaded area)

Note: Frequency of occurrence, *f*, is given for each circulation type.
* The ratio of mean daily precipitation of days of specific pattern to mean daily precipitation for all days.
Source: Steyn (1984).

coast trough-circulation type. The Durban–Bulawayo index gives a measure of the strength of the tropical easterlies; the Alexander–Windhoek index gives a measure of easterly waves and/or lows, and so on. Each of these indices, cumulated from daily data to give an annual value, may be correlated with yearly rainfall to produce rainfall correlation fields. The results are striking (Figure 18.9) and the distributions are in reasonable accord with results for monthly and daily scales.

Principal component analysis of the indices confirms the patterns found above (Tyson 1984). If the results of such analyses at 850 hPa and 500 hPa are linked to rainfall by a stepwise multiple correlation analysis, it is possible to isolate, for every rainfall station, not only the most important index, and hence circulation type, but also the level at which the control by that type is strongest. The technique produces highly coherent regions over which a particular circulation type is, year after year, the dominant influence on rainfall (Figure 18.10). Thus over northern central areas tropical easterlies and their perturbations are the single most important controls of the annual rainfall climatology. Over eastern South Africa the ridging anticyclones are dominant; western areas are affected most by west coast troughs, and so on. Most significantly, it turns out that, with the exception of a region to the far north-east, the annual average circulation at 500 hPa is a more important determinant of annual rainfall than the circulation near the surface.

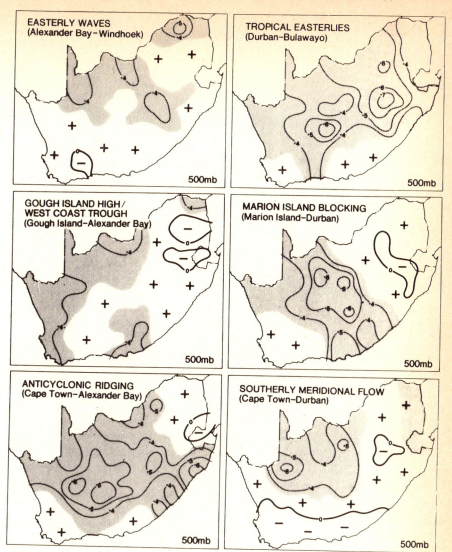

Figure 18.9 Examples of fields of correlation between individual 500 hPa station-pair geopotential indices (and hence geostrophic winds) and rainfall series

Note: Shading indicates regions in which correlation coefficients are locally significant at the 90 per cent confidence level. Positive signs denote regions of positive correlation; negative signs regions of negative correlation

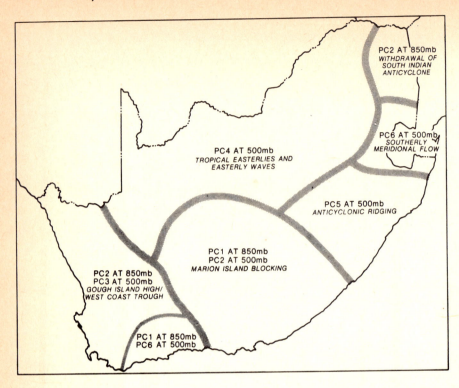

Figure 18.10   The regional distribution of predominant *annual* atmospheric controls of wet and dry spells

Note: Each region encompasses an area within which like circulation variables correlate most strongly with annual rainfall totals. The circulation variables have been defined by specific principal components at either 850 or 500 hPa and the strongest correlations have been determined by stepwise multiple correlation analysis of rainfall and principal components at the 850 and 500 hPa levels taken together. Each region delimits the area within which the same principal component circulation type appears as the first correlation in a stepwise multiple correlation analysis of rainfall and all principal components at 850 and 500 hPa.
Source: Tyson (1984).

Conclusion

From day to day both the type of synoptic disturbance and its location in relation to the subcontinent of southern Africa greatly affect the likelihood of rainfall. Anticyclones, if situated over the land, produce subsidence, inhibit rainfall and allow fine weather to prevail. Tropical disturbances take the form of easterly waves and lows (and very occasionally tropical cyclones). The locus of an individual wave trough controls the region likely to receive rainfall to the east of the axis of lowest pressure. Temperate disturbances result from perturbations in the westerlies. Similarly, the position of the trough line determines where rain will occur to the west of the trough. In all cases rain will occur only if the upper-level flow field and divergence is conducive to such an occurrence. Whether or not this will be the case depends on the position of the upper-level trough.

Just as the positioning of upper-level troughs affects daily rainfall, so on the scale of months, seasons, years and spells of years any mechanism changing the preferential location of upper-level troughs and ridges immediately modulates variations in the rainfall climatology of southern Africa. Changes in the loci of upper troughs have significant consequences for climatic change in the medium term and changing wave symmetries do much to determine individual wet and dry years as well as extended wet and dry spells.

Acknowledgements

The author wishes to thank Mr P. Stickler and Mrs W. Job for drawing the diagrams and the Foundation for Research Development of the CSIR and the University of the Witwatersrand, Johannesburg for funding the research.

References

Diab, R.D., Garstang, M. 1985. 'Wind power dependence upon weather systems', *International Journal of Ambient Energy* 6, 89–100.

Erasmus, D.A. 1980. 'The formulation of a classification procedure for specific use on cumulus cloud weather modification experiments', unpublished MSc thesis, University of Cape Town.

Garstang, M., Emmitt, G.D., Kelbe, B. 1981. *Rain augmentation in Nelspruit (RAIN), Final Report by Simpson Weather Assocs., Inc. to Water Research Commission*, Pretoria, Department of Water Affairs.

Gouws, V.C. 1963. 'Heavy rains over the eastern Cape during March 1963', *South African Weather Bureau Newsletter* 169, 60–3.

Harrison, M.S.J. 1984. 'A generalised classification of South African summer rain-bearing synoptic systems', *J. Climatol.* 4, 547–60.

Harrison, M.S.J. 1986. 'A synoptic climatology of South African rainfall variations', unpublished PhD thesis, University of the Witwatersrand.

Longley, R.W. 1976. *Weather and Weather maps of South Africa*, South African Weather Bureau Technical Paper, no. 3.

Lund, I.A. 1963. 'Map-pattern classification by statistical methods', *J. Appl. Met.* 2, 56–65.

McGee, O.S. 1974. 'The transport of water vapour over South Africa with different airflow types', *S. African Geogr.* 4, 290–6.

Miron, O., Tyson, P.D. 1984. 'Wet and dry conditions and pressure anomaly fields over South Africa and the adjacent oceans, 1963–1979', *Mon. Wea. Rev.* 112, 2127–32.

Steyn, P.C.L. 1984. *The relationship between the 300 hPa circulation pattern and rainfall classification in the Bethlehem region. Bethlehem Precipitation Research Project, Progress Report, 24*, Pretoria, South African Weather Bureau.

Taljaard, J.J. 1958. 'South African air-masses: their properties, movement and associated weather', unpublished Phd thesis, University of the Witwatersrand.

Taljaard J.J. 1981. 'The anomalous climate and weather systems of January to March 1974', South African Weather Bureau Technical Paper no. 9.

Triegaardt, D.O., Kits, A. 1963. 'Die drukveld by verskillende vlakke oor suidelike Afrika en aangrensende oseane tydens vyfdaagse reen-en droe periodes in suid-Transvaal en noord-Vrystaat gedurende die 1960–1961 somer', *South African Weather Bureau Newsletter* 168, 37–43.

Tyson, P.D. 1984. 'The atmospheric modulation of extended wet and dry spells over South Africa, 1958–1978, *J. Climatol.* 4, 621–35.

Tyson, P.D. 1986. *Climatic Change and Variability in Southern Africa*, Cape Town, Oxford University Press.

Vowinckel, E. 1955. 'Southern hemisphere weather map analysis: five-year mean pressures', *Notos* 4, 17–50.

Vowinckel, E. 1956. 'Ein Beitrag zur Witterungsklimatologie des suedlichen Mozambiquekanals', *Miscelañea Geofísica Publicada pelo Serviço Meteorológico de Angola em Comemoração do X Aniversário do Serviço Meteorológico Nacional*, Luanda, 63–86.

Part IV
Other Regions

Chapter 19

Inter-annual variations of tropical rainfall and upper circulations

J. Jacobeit   University of Augsburg
Federal Republic of Germany

Introduction

One of the most important climatic parameters of lower latitudes is the amount of rainfall. Its variations and anomalies which greatly influence natural and social systems of affected regions are caused by corresponding variations of atmospheric circulation. The great variety of circulation patterns implies that different regions are affected by varying connections of anomalies. Some of these connections may occur more frequently than other ones thus constituting a dominating link between quite different regions. The main aim of this chapter is, however, to consider some of those patterns which cause different connections of anomalies on the inter-annual time-scale; as an example we have chosen the northern hemispheric tropics between the Atlantic Ocean and the Bay of Bengal which include the drought-prone regions of the north African Sahel and the Indian subcontinent.

Rainfall indices

Monthly rainfall amounts of more than 400 stations within the global tropics were taken from Monthly Climatic Data for the World, covering the period 1968–85. Rainfall departures (i.e. differences of actual and mean (1968–85) monthly rainfall) must not be used for widely spaced comparison since their standard deviations are positively correlated with the absolute rainfall levels (e.g. with the means of annual precipitation). In order to achieve comparable quantities of rainfall departures even on a global scale, some kind of adjustment is needed. Most commonly used is the standardising procedure (i.e. dividing rainfall departures by the standard deviation of all these departures for each station) which, however, also equalises different quantities of rainfall variations at different stations. Dividing rainfall departures (DP) by the mean annual precipitation (MAP) of each station results in quantities which equalise the different absolute levels of rainfall at different stations and maintain their different levels of rainfall variation; however, this simple transformation, carried out with scaling factors (SF) of 100, yields percentages of DP with respect to MAP, the standard deviations of which are negatively correlated with the means of annual precipitation. Thus a middle course between departures and percentages defines the required rainfall index (RFI) if its standard deviations are uncorrelated with the means of annual precipitation:

$$ \text{RFI} = \text{DP} \cdot \left\{ \left(\frac{\text{SF}}{\text{MAP}} - 1 \right) \cdot f\,(\text{SF}) + 1 \right\} $$

The fractional factor f ($0 < f < 1$) depends on the scaling factor SF, and all couples of SF and f (SF) which achieve this non-correlation satisfy a hyperbolic equation which strongly approximates a linear one:

$$f (\text{SF}) \approx a \cdot \text{SF} + b$$

The line parameters a and b only vary significantly if the total number of stations is altered; this means that s_{RFI} and MAP remain uncorrelated. With SF = 1700 the standard deviation of all indices of all stations gives a value near 100 which also varies only a little if the total number of stations is altered.

In order to characterise a whole subregion its mean value $\overline{\text{RFI}}$ has been modified with respect to the general 'nature' of the averaged stations and to the actual quality of averaging. The standard deviation of the spatially averaged index $\overline{\text{RFI}}$ includes a factor ϱ which represents the dependence upon the number of averaged stations N and upon their mean rainfall correlation $\bar{r}$ (see Katz and Glantz 1986):

$$\varrho = \sqrt{\left(\frac{1}{N} + \left(1 - \frac{1}{N} \right) \cdot \bar{r} \right)}$$

The inverted relative variability v with respect to $\overline{\text{RFI}}$ (i.e. the mean of RFI deviations from $\overline{\text{RFI}}$ related to $\overline{\text{RFI}}$) is being weighted with ϱ for $v > 1$ and with $1 - \varrho$ for $v \le 1$ (low values of v are expected for strongly correlated stations, high values of v for weakly correlated ones). Considering the invariance of $\overline{\text{RFI}}/\varrho$ for $v = 1$ we get the subregion index SRI:

$$\text{SRI} \approx \begin{cases} \dfrac{\overline{\text{RFI}}}{\varrho} \cdot \left(\varrho + \dfrac{1 - \varrho}{v} \right) & \text{for } v \le 1 \\[3mm] \dfrac{\overline{\text{RFI}}}{\varrho} \cdot \left(1 - \varrho + \dfrac{\varrho}{v} \right) & \text{for } v > 1 \end{cases}$$

This rainfall index may be used for comparisons even on a global scale since it compensates for different rainfall levels and for different station groupings; at the same time it gives substantial characterisation of different subregions since it considers different orders of rainfall variations and different intra-regional spreadings of rainfall anomalies.

Rainfall anomalies

The subregional rainfall anomaly index has been applied not only to individual months but also to 'spells' of at least two months' duration or longer. Such spells are determined on the basis of monthly summarised RFI indices of equal sign related to the total number of stations within the subregion; if one of these sums exceeds the $s_{\text{RFI}}/2$ threshold for at least two successive months, the calculation of SRI is performed on the basis of RFI means for each station with respect to this actual spell. Thus the short-term fluctuations of individual months are distinguished from the more persistent deviations, as well as counteracting tendencies within

Table 19.1 Subregional rainfall anomaly index (SRI) for five Sahelian and five Indian subregions (see text) during significant spells of four summer monsoon seasons

	North African Sahel					Indian subcontinent				
	W	W/C	C	C/E	E	PN	N	M	S	SE
1972	−25.7	−85.8	−161.9	−135.0	−9.5	−138.6	−138.7	−137.6	−93.3	−99.4
1974	−44.0	58.8	194.4	199.0	99.9	−79.7	−193.4	−146.3	0.1	−50.6
1975	261.5	15.2	6.6	66.0	−137.9	333.7	211.8	23.5	172.9	58.6
1983	−174.8	−76.5	−131.7	–	−47.2	234.9	140.0	116.6	84.0	92.3

parts of the subregion or parts of the overall spell which are considered inherently.

The delimitation of subregions has been done according to the following criteria: i) equal course of mean monthly rainfall within the year; ii) similar coefficients of correlation between all pairs of stations; iii) additional criteria like mean annual precipitation, latitudinal and longitudinal extension, to ensure subregional dimensions of comparable order. Thus the north African Sahel (slightly differing from Nicholson 1979) has been delimited towards lower latitudes by means of the 500 mm isohyet and subdivided into five longitudinal regions: west (W, nine stations), up to 10 °W; west/central (W/C ten stations), 10 °W–4 °E; central (C, seven stations), 4–14 °E; central/east (C/E, four stations), 14–27 °E; east (E, three stations), east of 27 °E. Five subregions too, may serve for comparison with the Indian subcontinent between 67° and 85 °E: northern Pakistan and Kashmir (PN, five stations); eastern Pakistan and north India up to 26 °N (N, thirteen stations); middle India between 26° and 20 °N (M, eight stations); southern India except the south-east (S, nine stations); south-eastern India (SE, four stations). Table 19.1 shows the SRI index of these ten subregions for the significant anomaly spells during four different monsoon seasons (these spells have persisted between two and four months and have been determined as described above). Considering the $s/2$ threshold of significance of SRI ≈ 50 the different connections of anomalies between these continental regions emerge clearly. The year 1972, being part of the well-known Sahelian drought period of 1968–73, also shows negative deviations all over the Indian subregions. The year 1974, with further deficient monsoon months in India outside the southern part, has brought above average summer rainfall to the Sahel regions east of Senegal. The year 1975 represents a widespread wet monsoon season with roughly normal conditions in middle India and the central Sahel (only the few stations in the outermost east suffered from dryness). The year 1983, with further wet spells all over the Indian subregions, again shows remarkable Sahelian deficits which only in the eastern part do not reach entirely the threshold of significance.

Upper circulations

To understand the atmospheric circulation patterns which run parallel to these different connections of tropical rainfall anomalies, we have used the daily National Center for Atmospheric Research (NCAR) sets of horizontal wind components at the 200 and 700 hPa level which are given for a tropical grid with 5° of longitudinal resolution and (according to constant distances on a Mercator map) decreasing

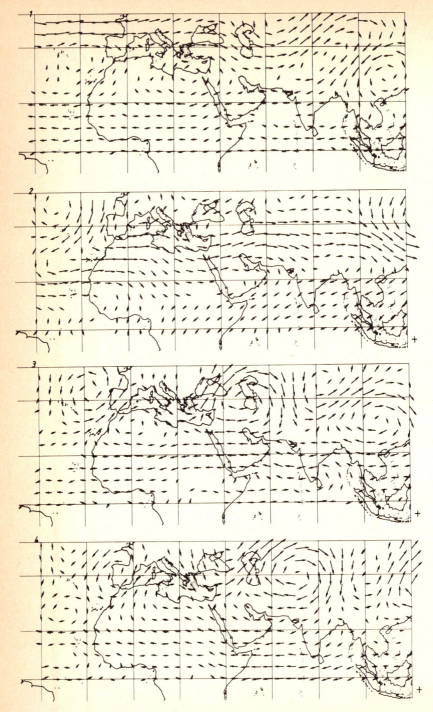

Figure 19.1   Principal components 1–4 of the summer monsoonal 200 hPa flow of 1972, 1974, 1975 and 1983

Table 19.2   Seasonal percentages of variance explained by the four principal
components of Figure 19.1

Component	1	2	3	4
72:	20.8	28.1	11.9	6.4
74:	29.9	19.0	14.0	6.2
75:	19.3	12.7	32.2	10.0
83:	11.9	21.5	18.2	19.0

differences with latitude. For the present purpose we have selected the section
between the equator and 48.1 °N and between 60 °W and 120 °E. Referring to the
above-mentioned summer monsoon anomalies, we have further restricted the data
sets to the 100-day period from 8 June to 15 September. The successive four-day
averages from these periods in 1972, 1974, 1975 and 1983 were submitted to
principal component analysis from which the first four components referring to the
200 hPa level are reproduced in Figure 19.1. These components represent independent tendencies of the upper tropospheric circulation during these four different
seasons and explain some 70 per cent of the total variance. The percentage of the
variance of the separate seasons explained by these components is shown in Table
19.2. From this it can be seen that component 1 is represented much more in 1974
than in 1983, component 2 much more in 1975; component 3 reaches its highest
value in 1975, component 4 in 1983. These differences give rise to some conclusions concerning the components' meaning which will be complemented by some
other circulation variables. Figure 19.2 shows approximations of the mean seasonal
vertical movement which is evaluated from the difference between the mean
seasonal horizontal divergence at the 200 and at the 700 hPa level. These
divergences, given in cartesian coordinates by

$$\text{div} = \frac{\partial u}{\partial x} + \frac{\partial v}{\partial y}$$

(where u and v are horizontal wind components), are estimated from the differences
of the zonal component in zonal direction and of the meridional component in meridional direction. Figure 19.3 shows mean seasonal deviations of the lower
tropospheric 700 hPa vertical component of the relative vorticity from its four-
seasonal mean which may be considered as significant beyond the actual level of
approximately ± 2.0. The relative vorticity, given in cartesian coordinates by

$$\zeta = \frac{\partial v}{\partial x} - \frac{\partial u}{\partial y},$$

has been estimated from the differences of the meridional wind component in zonal
direction and of the zonal wind component in meridional direction.

Results

The widespread monsoonal season of 1972 is characterised by an enhanced
representation of the 200 hPa patterns with westerly flow reaching far towards the

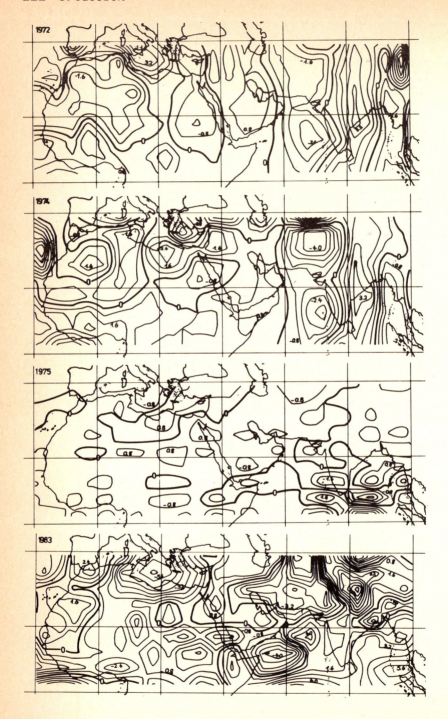

Figure 19.2 Differences in mean seasonal horizontal divergence at the 200 and 700 hPa level

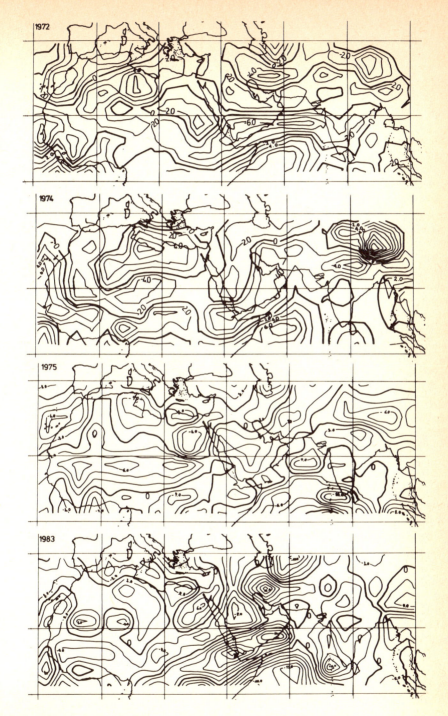

Figure 19.3 Mean seasonal deviations in the relative vorticity at the 700 hPa level from its four-seasonal mean

equator (down to 28 °N above India and even 20 °N above Africa) and substantially weakened tropical easterlies (component 2). The 'Tibetan' high, which is decisive for the strength and effectiveness of the easterlies in the longitudinal range, seems very weak and far displaced south-eastwards. These features fit with former studies about dry monsoonal seasons (Ramaswamy 1962; Kanamitsu and Krishnamurti 1978). A marked northerly component within the disturbed easterlies leads to widespread cross-equatorial flow which seems to be most extensive around Africa. The high-amplitude trough off the north-west-African coast (which is reflected in the mean seasonal upward movement and the positive deviations of the mean 700 hPa vorticity there) enforces the southerly path of the westerlies upstream from which another cross-equatorial component is separated by anticyclonic turns above west Africa. This is still reflected at the 700 hPa level where negative deviations of the mean vorticity extend far towards the equator. Only above the eastern regions are some positive deviations to be seen where the upper flow runs through a weak cyclonic wave. Above the Arabian Sea another centre of negative deviations developed below the upper anticyclonic turns and marginally extends beyond the Indian subcontinent. Correspondingly, the mean vertical movement is directed downward above the Arabian Sea with the neutral line passing through Western India. Above Africa there is widespread sinking in the mean which extends far southward into the central Sahelian dry region.

These conditions, which favour below average summer monsoon rainfall in most of the marginal-tropical regions of our section, are represented above the average in 1972 and below the average in the widespread wet season of 1975 (See Table 19.2). There, in contrast, other tendencies prevail (component 3): the upper tropical easterlies are well defined and reach to significantly more northerly regions (up to 24 °N outside of cellular influenced regions). The Saharan anticyclone also lies in a northern position and does not effect any remarkable extension of negative vorticity deviation towards lower latitudes. Upper cross-equatorial flow is only indicated above Africa, but with a northerly direction. Eastwards, there is westerly flow near the equator which turns cyclonically around southern India. This is reflected in mean upward motions south of India which, however, are inverted from the southern continental regions onwards. The small mean vertical movement above Africa, too, does not indicate any peculiarity but hints at greater intra-seasonal differences which also may appear in neutral SRI values of the central Sahelian subregions. The mean deviations of 700 hPa vorticity, however, show widespread positive values above tropical Africa, especially above the Sahelian regions, and hint at dominating lower-level influences in generating the actual wet spells. The positive deviations above the Arabian Sea and north of India seem to have an analogous effect on monsoonal rains in northern India as cyclonic curvatures aloft in the southern part.

The opposite tendencies of 1972 and 1975 imply similar deviations in monsoonal rainfall above most of India and the Sahelian regions. There are, however, other patterns too, which lead to different deviations in these two regions. Component 1 of 200 hPa flow, for example, which is equally represented in the opposite years of 1972 and 1975, is strengthened during 1974 (wet Sahelian and dry Indian spells) and weakened during 1983 (dry Sahelian and wet Indian spells). It is characterised by well-developed easterlies above Africa, whereas the westerly trough north of the Arabian Sea has pushed the anticyclone downstream with major gaps within the easterlies above the Bay of Bengal and the Arabian Sea and continuously developed easterlies only farther south. In 1974 this leads to negative deviations in the mean

700 hPa vorticity north of India with slowly rising values down to the south; additionally the distribution of mean upward/downward motion closely resembles the 1972 pattern. Above the Sahara, however, the lower-level anticyclone seems distinctly weakened (positive vorticity deviations) with upward motions above north-western Africa which may be triggered by the abating winds between the upper anticyclonic cells to the west and to the east and may exert a stimulating influence upon the African monsoonal system.

The opposite season of 1983 additionally shows a greater representation of component 4 which is characterised by a similar wave pattern to component 3 but shifted eastward some 15–20° of longitude. This means that a strong anticyclonic cell is present north of India which accelerates the easterly 200 hPa flow above the Indian Ocean. Above north Africa its main current still turns farther north but loses kinetic energy and becomes involved into the circulation of the Saharan anticyclone. At lower levels negative deviations of the mean vorticity dominate south of 20 °N with widespread sinking motion in the mean that causes dry spells up to the Soudan belt. In contrast to that the Indian region is characterised by positive vorticity deviations and upward motions in the mean the core of which, however, lies more westwards above the Arabian Sea.

Conclusion

These examples demonstrate that different connections of tropical rainfall anomalies are linked with different circulation tendencies and patterns. Even if one or several of these connections dominate over long-term periods this may not necessarily hold for any other period: major transitions of climatic states, for example, may alter preferred tendencies and patterns of circulation thus inducing changed connections of anomalies.

References

Kanamitsu, M., Krishnamurti, T.N. 1978. 'Northern summer tropical circulations during drought and normal rainfall months', *Mon. Wea. Rev.* 106, 331–47.

Katz, R.W., Glantz, M.H. 1986. 'Anatomy of a Rainfall Index', *Mon. Wea. Rev.* 114, 764–71.

Nicholson, S.E. 1979. 'Revised rainfall series for the West African subtropics', *Mon. Wea. Rev.* 107, 620–3.

Nicholson, S.E. 1980. 'The nature of rainfall fluctuations in subtropical West Africa', *Mon. Wea. Rev.* 108, 473–87.

Ramaswamy, C. 1962. 'Breaks in the Indian summer monsoon as a phenomenon of interaction between the easterly and the subtropical westerly jet streams', *Tellus* 14, 337–49.

Chapter 20

El Niño years and the spatial pattern of drought over India, 1901–70

S. Gregory University of Sheffield, United Kingdom

El Niño and drought years

It has been argued by many writers (e.g., Bhalme and Jadhav 1984; Bhalme *et al.* 1983; Mooley and Parthasarathy 1983; Pant and Parthasarathy 1981; Parthasarathy and Pant 1985) that one critical factor controlling fluctuations of the Indian summer (June to September) monsoon rainfall is changes in the Southern Oscillation. Moreover, the relationship of the El Niño phenomenon in the eastern equatorial Pacific Ocean with the Southern Oscillation (the ENSO phenomenon) is then developed to suggest a strong link between El Niño and widespread drought in India over this four-month season. That this relationship is far from universal even in terms of India as a whole must be recognised, however. Not only are all El Niño years not drought years, but also many drought years occur in non-El Niño years.

Using the definition of El Niño years proposed by Rasmussen and Carpenter (1983), there were 16 such events within the 70 years 1901–70. In Table 20.1 the extent of summer monsoon drought in each of these years is suggested by four indices. Clearly, 1918 was the year of most widespread summer monsoon drought or rainfall-deficiency conditions by all four criteria, and was also the year of most intensive El Niño conditions. The other five years of strongest El Niño condition all occur in the rankings on all four indices, but not in the same order. Moreover, many of the years with rather weaker El Niño conditions did not lead to sufficiently widespread summer monsoon drought to be ranked as years of marked rainfall deficit. In addition, as gaps in the ranking imply, several years without El Niño conditions experienced widespread drought in this season. In Table 20.2 are listed those years without El Niño where there were either at least 100 stations with drought by Method 1 or at least 50 stations by Method 2 (using the same four indices as in Table 20.1).

Of these years without El Niño conditions, it was 1920 in which summer monsoon drought conditions were most widespread., this being within the most extreme four years by all the criteria used. It should also be stressed that in each of the 70 years there were always some of the 306 stations at which drought in this season occurred by both Method 1 and Method 2. The fewest such stations were in 1917 (5/2), 1910 (15/3) and 1964 (16/4), but there were only 8 of the 54 non-El Niño years which had fewer than the 25 stations with drought by Method 1 that occurred in the El Niño years of 1914 and 1953.

Spatial patterns of drought in El Niño years

From this general background, the main objective of this chapter is to consider the spatial pattern of summer monsoon drought conditions to assess: (a) whether there

Table 20.1   Four indices of the extent of summer monsoon (June to September) drought conditions in India in the El Niño years[1] within the period 1901–70, ranked in the context of all 70 years

Year	Method 1[3]	Method 2[4]	Method 3[5]	Method 4[6]
1902	63	24	1	
1905[2]	148 (2)	99 (2)	10 (2.5)	(5)
1911[2]	120 (5.5)	71 (5)	10 (2.5)	(6)
1914	25	13	0	
1918[2]	212 (1)	175 (1)	19 (1)	(1)
1923	73	32	5	
1925	87	38	4	
1930	56	18	2	
1932	77	30	1	
1939	100 (14)	43 (17)	6 (11.5)	
1941[2]	120 (5.5)	74 (3.5)	7 (9.5)	(8)
1951[2]	117 (8)	63 (11)	5	(2)
1953	25	15	0	
1957	58	28	2	
1965[2]	124 (4)	64 (10)	8 (7)	(3)
1969	53	26	2	

Notes: 1. El Niño years from Rasmussen and Carpenter (1983), when the East Equatorial sea surface temperature was $\geq$ + 1.0 °C above average, and the change from one year to the next was $\geq$ + 2.5 °C
   2. The most marked El Niño years
   3. Method 1: the number of district stations out of 306 with rainfall < 75% of the 1901–70 average (rank in brackets)
   4. Method 2: the number of district stations out of 306 where the year was one of the driest 10% in the period 1901–70 (rank in brackets)
   5. Method 3: the number of Meteorological Sub-divisions out of 29 where the year was one of the driest 10% in the period 1901–70 (rank in brackets)
   6. Method 4: the rank (in brackets) for those years in which the all-India data was < 90% of the average (from Bhalme and Jadhav, 1984)

Table 20.2   Four indices of the extent of widespread summer monsoon (June to September) drought conditions in India in critical non-El Niño years within the period 1901–70, ranked in the context of all 70 years

Year	Method 1	Method 2	Method 3	Method 4
1901	113 (9.5)	54 (14)	6 (11.5)	(7)
1904	113 (9.5)	69 (6.5)	8 (7)	(9)
1907	119 (7)	61 (12)	4	
1913	111 (11)	66 (9)	5	(11)
1915	101 (13)	69 (8)	9 (4.5)	
1920	125 (3)	74 (3.5)	9 (4.5)	
1928	107 (12)	69 (9)	5	
1952	95 (15.5)	53 (15)	7 (9)	
1966	95 (15.5)	60 (13)	8 (7)	(10)

* For the definition of Methods 1, 2, 3, 4 see Table 20.1.

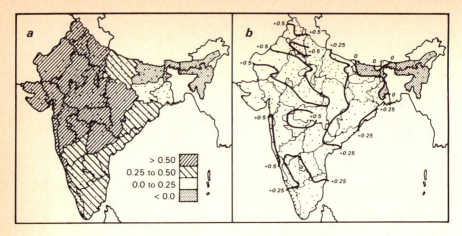

(a) 29 Standard Meteorological Sub-divisions
(b) 306 District Stations

Figure 20.1   Correlation coefficients for the period 1901–70 between the all-India summer monsoonal rainfall data set and local data

are distinctively different patterns in different years; (b) if so, are such patterns different between El Niño and non-El Niño years; (c) if similar patterns are shared between the two sets of years, whether there are nevertheless El Niño tendencies in terms of such drought spatial patterns; and (d) whether there are some areas of India where El Niño-related summer monsoon droughts are frequent and other areas where El Niño conditions provide no more than a limited indication of probable drought conditions.

Many earlier studies have considered summer monsoon drought and flood conditions in India in terms of an integrated overall all-India data set, reflecting the total amount of rain falling over India as a whole during the period June to September (e.g. Bhalme and Mooley 1980; Mooley *et al.* 1981; Mooley and Parthasarathy 1983; 1984), or in terms of a division into northern, central and southern India (Ananthakrishnan and Parthasarathy 1984). To assess spatial patterns, however, it is necessary to use, for example, either the 29 Meteorological Sub-divisions or the 306 District stations mentioned in Tables 20.1 and 20.2 (Gregory 1986).

The results from using either of these will be similar in general, although there will be detailed differences especially in terms of the magnitude of any values or impacts. However, temporal relationships established between El Niño conditions and the widely used all-India data set will not apply equally over the whole of the country at these more detailed levels, and in some areas they will be very misleading. Thus, as the maps of correlation between local data and the all-India data show in Figure 20.1, coefficients exceed + 0.5 only in the north-western sector of the country, and they become weakly negative in the north-east (Gregory 1986).

To investigate whether there is any common spatial pattern or patterns during El Niño years, these 16 years of summer monsoon rainfall were subjected to a principal components analysis, having first reduced the number of stations to 240

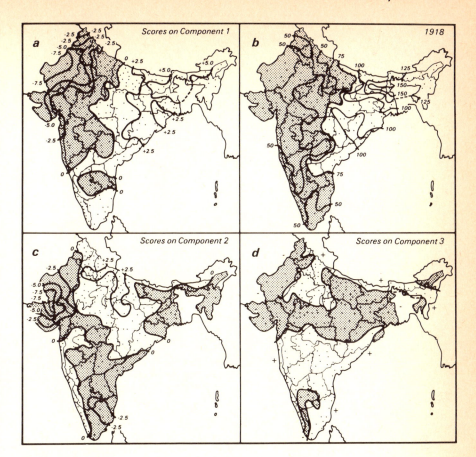

(a) Component 1[1]
(b) Summer monsoon rainfall of 1918 as % of the 1901–70 average, for 306 stations[2]
(c) Component 2[1]
(d) Component 3[1]

Figure 20.2   Scores derived from a direct principal components analysis of summer monsoonal rainfall for 16 El Niño years within the period 1901–70 over 240 stations (a), (c), (d) and summer monsoonal rainfall of 1918 as a percentage of the 1901–70 average for 306 stations (b)

1. Negative values stippled
2. Areas with drought deficits of more than 25% are stippled

to prevent over-representation of areas with a higher station density. The first component determined no more than 19 per cent of the variance and the second only 14 per cent whilst each of the first eight components contained the highest loading for at least one of the 16 El Niño years. The assumption of one or two critically similar patterns, leading to a few components each determining a large proportion of the variance, thus does not apply within these years. Of the six years defined in terms of the all-India data (Method 4) in Table 20.1, 1905, 1911 and 1918 (together with 1939) loaded heavily on component 1, the respective loadings being

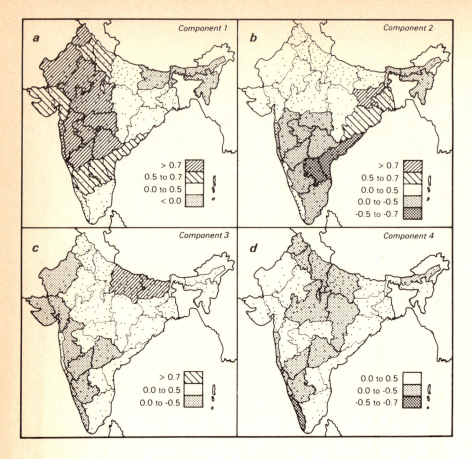

(a) Component 1
(b) Component 2
(c) Component 3
(d) Component 4

Figure 20.3   Loadings derived from a direct principal components analysis of summer monsoonal rainfall of 29 Standard Meteorological Sub-divisions for the 70 years 1901–70

+ 0.70, + 0.78, + 0.83 and + 0.72. In contrast, 1941 loaded heavily on component 2 (− 0.61), 1965 less heavily on component 3 (+ 0.51), and 1951 on component 5 (+ 0.50). Thus although these marked El Niño years were all characterised by widespread drought, the pattern varied, and apart from the situation on component 1 none of the other heaviest loadings accounted for as much as 50 per cent of the variance. Moreover, apart from the four years mentioned earlier, none of the other 12 El Niño years were strongly loaded on component 1, and in fact nine of those years had loadings of between only − 0.15 and + 0.25, i.e. the spatial patterns of those years were clearly different from that of component 1. The pattern itself for component 1 is shown in Figure 20.2a, and as an illustration the conditions for 1918 are presented in Figure 20.2b. In addition, the different patterns for components 2 and 3 are shown in Figures 20.2c and 20.2d respectively.

The spatial patterns of drought in all years

The existence of one strong spatial pattern (component 1) within El Niño years, accompanied by a number of other different though less well-developed patterns, is thus clear. Such tendencies may, however, also apply to non-El Niño years. To examine this a similar principal components analysis was carried out on all 70 years together, using the Meteorological Sub-division data from Parthasarathy *et al.* (1987). This necessarily yields a more generalised result, but the basic spatial patterns can be seen.

The spatial pattern for component 1 in this analysis (Figure 20.3a) is effectively the same as that of the earlier analysis simply on El Niño years. This time, however, approximately 30 per cent of the total variance is determined, which suggests the development of such conditions with a considerable degree of frequency. If the standardised scores are then plotted as a time series (Figure 20.4a) it can be seen that of the 12 years with a value of below − 1.0, i.e. with a marked summer monsoon rainfall deficit, seven are El Niño years and five are not. Clearly this is a common drought pattern which, although more typical of El Niño years, is nevertheless also possible under other circumstances. In the years when this pattern was inverted, however, it was only 1914 that was an El Niño year, but the occurrence of wet conditions in the north-west and dry conditions in the north-east in any El Niño year is worthy of comment.

The pattern for component 2 (Figure 20.3b), which comprises 12.5 per cent of the total variance, is comparable to component 3 when only El Niño years were considered. It can be seen from the standardised scores (Figure 20.4b) that 1965 summer monsoon rainfall deficits strongly follow this pattern. Apart from 1918, the other nine years with this pattern of deficits were all non-El Niño years, as were all ten of the years with a reverse pattern, i.e. southern surpluses and northern deficits. Similar patterns and temporal relationships are also present for components 3 (only 8 per cent of the total variance) and 4 (only 6.5 per cent of the total variance) in Figures 20.3c and 20.3d, and Figures 20.4c and 20.4d respectively.

General review

Earlier, four questions were asked, and we can now indicate answers to three of them. That there are different and distinctive spatial patterns of drought and rainfall deficits in different years is clear from the maps and graphs presented. Also, it has been shown that of the several more common patterns, none of them is specific to El Niño years and absent from non-El Niño years, though the relative frequency of them may differ between the two sets of years. This implies that there are tendencies for certain patterns to occur in El Niño years, and this is especially true of years with strong El Niño conditions. Under these circumstances, the pattern displayed by component 1 in Figures 20.2a, 20.3a and 20.4a and illustrated by year 1918 in Figure 20.2b, is very common, though it is neither universal nor limited to El Niño years only. This pattern displays rainfall deficit conditions, often of

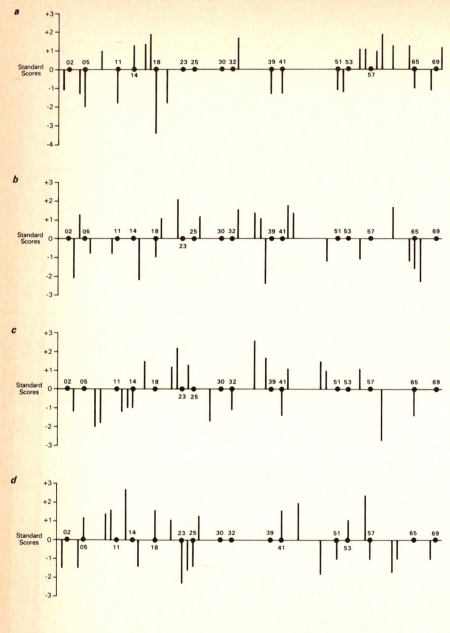

(a) Component 1
(b) Component 2
(c) Component 3
(d) Component 4

Figure 20.4 Standardised scores ($\geqslant +1.0$ and $\leqslant -1.0$) for the 70 years 1901–70, from a direct principal components analysis of the summer monsoonal rainfall of 29 Standard Meteorological Sub-divisions, with 16 El Niño years indicated by their dates

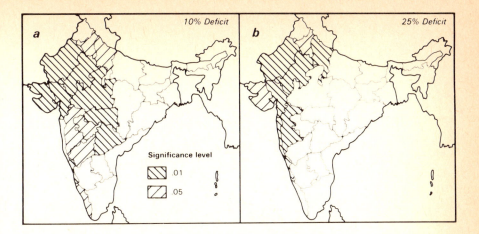

Figure 20.5   Areas where summer monsoonal rainfal deficits below the 1901–70 average of (a) ⩾ 10% (b) ⩾ 25% were significantly more frequent in El Niño years than in non-El Niño years, in terms of Fisher's Exact test

drought intensity, in the western half of northern India, extending southwards over the Deccan and along the western coast, but with the above average rainfall over the eastern half of northern India and transitional values in the south of the country, apart from the western coast. Moreover, it is in these same western areas that the occurrence of rainfall deficits of more than 10 per cent during El Niño years, as compared to their occurrence in non-El Niño years, is significantly different from random in terms of Fisher's Exact test (Figure 20.5a). For more extreme deficits of drought intensity (i.e. more than 25 per cent) the area where El Niño occurrences are clearly more likely to be related to these conditions than are non-El Niño occurrences is more restricted (Figure 20.5b).

The fourth question concerned the definition of areas where El Niño deficits and droughts could be shown to be common, and other areas where the absence of such droughts may be the norm. The observed probabilities (in the period 1901–70) of deficits of more than 10 per cent in summer monsoon rainfall in strong El Niño years and also in all El Niño years, and then similarly for deficits greater than 25 per cent (i.e. droughts), are categorised into low, medium and high probabilities (Table 20.3).

It would be possible to construct a map for each of the columns in Table 20.3, but it is more useful to integrate these four sets of low, medium and high grades into one map to display the regional contrasts that exist. In this way the fourth question raised earlier can be resolved. In Figure 20.6a seven categories have been defined. This has involved certain generalisations to ensure that each category had either all four terms in the same grade or two in one grade and two in another. This has required adjusting, when necessary, no more than one of the four grades to its next nearest grade, to permit the restriction of the result to simply seven categories. These are indicated by suffix 2 in Table 20.3.

In the north-west (Harayana, Punjab, West and East Rajasthan, Gujarat, Saurashtra and Kutch) there is a high probability of both moderate and large deficits, both for the strong El Niño years and for all El Niño years (category 1).

Table 20.3   The probability of experiencing moderate (> 10%) or large (> 25%) deficits in summer monsoon rainfall in years of strong El Niño and in all El Niño years (over the period 1901–70), in terms of meteorological sub-divisions[1]

El Niño status	Moderate Deficit (> 10%)		Large Deficit (> 25%)	
	Strong	Any	Strong	Any
High probability	⩾ 5	⩾ 10	⩾ 4	⩾ 6
Medium probability	3 to 4	7 to 9	2 to 3	2 to 5
Low probability	⩽ 2	⩽ 6	⩽ 1	⩽ 1
03 N. Assam	L	L	L	L
04 S. Assam	L	L	L	L
05 Sub-Himalayan W. Bengal	L	L	L	L
06 Gangetic W. Bengal	L	L	L	L
07 Orissa	L	L	L	L
08 Bihar Plateau	L	L	L	L
09 Bihar Plains	L	L	L	L
10 E. Uttar Pradesh	M	M[2]	M	M
11 W. Uttar Pradesh Plains	H	M	H	M
13 Harayana	H	H[2]	H	H
14 Punjab	H	H[2]	H	H
17 W. Rajasthan	H	H	H	H
18 E. Rajasthan	H	H	H	H
19 W. Madhya Pradesh	M[2]	M	M	M
20 E. Madhya Pradesh	M	M[2]	M	M
21 Gujarat	H	H	H	H
22 Saurashtra and Kutch	H[2]	H	H	H
23 Konkan	H	M[2]	H	M
24 Madhya Maharashtra	H	M	H	M
25 Marathwada	M[2]	M	M	M
26 Vidarbha	H	H	M	M
27 Coastal Andhra Pradesh	M	M	L	L
28 Telengana	M[2]	M	M	M
29 Rayalseema	M	M	M	M
30 Tamil Nadu	L	M	L	M
31 Coastal Karnataka	M[2]	M	M	M
32 N. Karnataka	M	M	M	M
33 S. Karantaka	M	M[2]	M	M
34 Kerala	M[2]	M	L	L

Notes: 1. The key number of meteorological sub-divisions are shown on Figure 20.6a

2. These grades have been adjusted for the regionalisation in Figure 20.6a

In contrast, in the north-east (North and South Assam, sub-Himalayan and Gangetic West Bengal, Orissa, Bihar Plateau and Bihar Plains) probabilities are low in all four sets of conditions (category 7) and in many cases they were zero during the observation period. Between these two areas in the north, and extending southwards over most of the Deccan, probabilities are medium in all four sets of conditions (category 4), although in many cases one of these sets has been adjusted (see Table 20.3). In West Uttar Pradesh Plains, Konkan and Madhya Maharashtra, however, probabilities of both moderate and large deficits are high under strong El Niño conditions but only medium under overall El Niño conditions (category 3), whilst in Vidarbha they are high for moderate rainfall deficits but only medium for

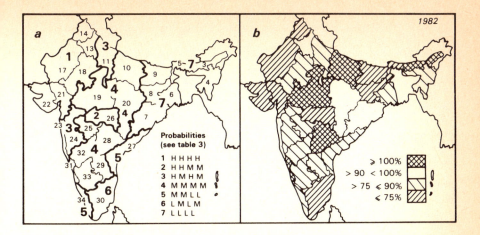

Figure 20.6 (a) Regional pattern based on integrating low, medium and high observed probabilities (1901–70) of moderate and large summer monsoonal rainfall deficits, under strong and all El Niño years; (b) Sub-division summer monsoonal rainfall of 1982 expressed as a percentage of the 1901–70 average

Note: For definitions see Table 20.3 (a).

large deficits of more than 25 per cent (category 2). Again, in both coastal Andhra Pradesh and Kerala moderate deficits of more than 10 per cent have a medium probability whatever El Niño conditions are considered, but deficits of more than 25 per cent occur with only a low probability even in strong El Niño years (category 5). Finally, in Tamil Nadu both sets of deficits show only a low probability of occurrence when strong El Niño conditions exist, but medium probability in relation to the broad range of El Niño conditions (category 6), suggesting that it is in the years of a weaker rather than a stronger El Niño that summer monsoon rainfall deficits are more likely to occur in Tamil Nadu.

That these are only observed probabilities based on the events of the 1901–70 period can be illustrated by the later strong El Niño year of 1982. Subdivision rainfall for this year, as a percentage of the 1901–1970 average, is mapped in Figure 20.6b. This displays the less common spatial pattern of component 2 of Figure 20.2c rather than that of the more common component 1 of Figure 20.2a, and 1982 was thus more comparable to 1941 rather than 1905, 1911 or 1918. As a result, it was in the areas of normally low probability of moderate or large deficits during strong El Niño years (categories 6 and 7 in Figure 20.6a) that 1982 proved to be most unusual.

References

Ananthakrishnan, A., Parthasarathy, B. 1984. 'Indian rainfall in relation to the sunspot cycle: 1871–1978', *J. Climatol.* 4, 149–69.

Bhalme, H.N., Jadhav, S.K. 1984. 'The Southern Oscillation and its relation to monsoon rainfall', *J. Climatol.* 4, 509–20.

Bhalme, H.N., Jadhav, S.K., Mooley, D.A., Ramana Murthy, B. V. 1986. 'Forecasting of monsoon performance over India', *J. Climatol.* 6, 347–54.

Bhalme, H.N., Mooley, D.A. 1980. 'Large scale droughts/floods and monsoon circulation', *Mon. Wea. Rev.* 108, 1197–1211.

Bhalme, H.N., Mooley, D.A., Jadhav, S.K. 1983. 'Fluctuations in the drought/flood area over India and relationship with the Southern Oscillation', *Mon. Wea. Rev.* 111, 86–94.

Gregory, S. 1986. 'Data base aspects of temporal patterns of climatic fluctuations: an Indian case study', *Trans. Inst. Indian Geogr.* 8, 35-44.

Mooley, D.A., Parthasarathy, B. 1983. 'Variability of the Indian summer monsoon and tropical circulation features', *Mon. Wea. Rev.* 111, 967–75.

Mooley, D.A., Parthasarathy, B. 1984. 'Fluctuations in all-India summer monsoon rainfall during 1871–1978', *Climatic Change* 6, 287–301.

Mooley, D.A., Parthasarathy, B., Sontakke, N.A.,, Munot, A.A. 1981. 'Annual rainwater over India, its variability and impact on the economy', *J. Climatol.* 1, 167–86.

Pant, G.B., Parthasarathy, B. 1981. 'Some aspects of an association between the Southern Oscillation and Indian summer monsoon', *Arch. Met. Geoph. Biocl.* ser. B., 29, 245–52.

Parthasarathy, B., Pant, G.B. 1985. 'Seasonal relationships between Indian summer monsoon rainfall and the Southern Oscillation', *J. Climatol.* 5, 369–78.

Parthasarathy, B., Sontakke, N.A., Munot, A.A., Kothawale, D.R. 1987. 'Droughts/floods in the summer monsoon season over different Meteorological Subdivisions of India for the period 1871–1984', *J. Climatol.* 7, 57–70.

Rasmussen, E.N., Carpenter, T.H. 1983. 'The relationship between eastern equatorial Pacific sea-surface temperatures and rainfall over India and Sri Lanka', *Mon. Wea. Rev.* 111, 517–28.

Chapter 21

Predictability of seasonal rainfall in the northern Nordeste region of Brazil

M.N. Ward, S. Brooks and C.K. Folland
Meteorological Office, United Kingdom

Introduction

A recent activity of the Synoptic Climatology branch at the UK Meteorological Office has been to investigate the importance of large-scale anomalies in sea-surface temperature (SST) to low-frequency atmospheric variability in the tropics and the extra-tropics. Progress has been facilitated by the creation of a quality-controlled data set of monthly SSTs covering most of the globe away from the Southern Ocean. The data have been adjusted for changes in methods of measuring SST (Parker and Folland, Chapter 4, this volume). Several recent empirical and physically-based studies have related large-scale sea-surface temperature anomaly (SSTA) patterns to tropical atmospheric anomalies near to and remote from the anomalous SST development (for example, Lau and Oort 1985, Owen and Folland, Chapter 13, this volume). Here we report an empirical study of rainfall in the northern Nordeste region of Brazil (see Figure 21.1).

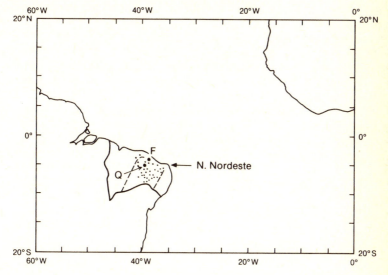

Key:

∴	Stations used to compile Hastenrath's northern Nordeste rainfall series
Q	Quixeramobim
F	Fortaleza
– – – –	Delineates area of northern Nordeste with annual rainfall less than 800 mm

Figure 21.1 Location map

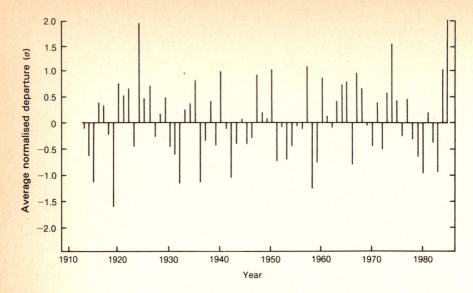

Figure 21.2 Hastenrath's standardised March–April rainfall series for the northern Nordeste, 1913–85

Rainfall in the northern Nordeste is usually concentrated in the months March to May (Namias 1972) when the inter-tropical convergence zone reaches its most southerly location (though still mostly to the north of the Nordeste). There is, however, substantial inter-annual variability in the seasonal rainfall (Figure 21.2). A number of authors have related inter-annual northern Nordeste rainfall fluctuations to large-scale SSTAs and atmospheric circulation changes. In particular, Hastenrath and Heller (1977) and Moura and Shukla (1981) established links between northern Nordeste rainfall and a north–south dipole pattern of SSTAs in the tropical Atlantic. The latter study used atmospheric general circulation model (AGCM) experiments to reinforce the empirical evidence. Earlier, Namias (1972) found a statistical relationship between northern Nordeste rainfall and preceding winter depression activity off Newfoundland. Weather events throughout the Atlantic sector may therefore be connected to northern Nordeste rainfall.

SST fluctuations in the Pacific may also be important (see, for example, Covey and Hastenrath 1978). El Niño events have coincided with a number of severe droughts in the northern Nordeste, and the Southern Oscillation Index is at times strongly associated with northern Nordeste rainfall, though the long-term correlation is quite low.

Hastenrath, Wu and Chu (1984) synthesised existing knowledge of teleconnections with northern Nordeste seasonal rainfall to produce a promising multiple regression forecasting system using both atmospheric and oceanic predictors. Later we develop and assess a forecast technique which uses large-scale patterns of SSTs as the sole set of predictors. This may seem restrictive but, in addition to their role as forcing agents, the SSTs partly reflect previous and simultaneous atmospheric states and have some advantage over atmospheric predictors because large scale SST patterns are less noisy. The SST data are believed to be reliable back to about

1900 (Folland, Parker and Palmer, 1986) and this enables the forecast experiments presented later to extend over a longer epoch than previously possible. SSTAs are also available in near real-time which made it possible to release an experimental forecast for the 1987 rainfall season in early March 1987.

Data and analysis

SST data and covariance SST eigenvectors

We use the extensively quality-controlled Meteorological Office Ocean Sea Temperature data set (Parker and Folland, Chapter 4, this volume) compiled from ship reports. One approach could have been to relate northern Nordeste rainfall to a set of predictors, each representing the SSTA in a particular area. However, information may be lost because it appears that the atmosphere can be most responsive to the overall pattern of large-scale anomalies (see, e.g. Palmer 1986). This suggests that SST predictors should measure the strength of large-scale spatial patterns in the SSTA field.

We have used eigenvector (empirical orthogonal function (EOF)) analysis to reveal some of the principle patterns of SST variability. Here we concentrate on SST patterns in the Atlantic and Pacific Oceans as these regions are found at present to be most relevant to predicting northern Nordeste rainfall. Atlantic and Pacific covariance EOFs have been calculated from 80 seasonal fields of SSTAs for the months of December to February for the period 1901–80 on a $10°$ latitude $\times$ $10°$ longitude space scale. The data have been interpolated where necessary using Chebychev polynomial time series. If the EOF analysis has been successful then a small set of EOF coefficient time series will trace the time variations of the principal large-scale Atlantic and Pacific SST patterns. Table 21.1 shows that this may be true, since (a) the first three Atlantic and four Pacific EOFs are statistically separated according to the test given by North et al. (1982) and (b) the EOFs are easily statistically distinguishable from noise according to Craddock's log eigenvalue test (Craddock and Flood 1969).

Table 21.1 Characteristics of first four Atlantic and Pacific December–February covariance EOFs*

| | Atlantic | | | Pacific | | |
	% of variance	Test A (95% level)	Test B	% of variance	Test A (95% level)	Test B
EOF						
1	14.5	Pass	Pass	14.9	Pass	Pass
2	8.9	Pass	Pass	11.0	Pass	Pass
3	7.0	Pass	Pass	5.2	Pass	Pass
4	6.7	Fail	Pass	4.2	Pass	Pass
Total % of variance	37.1			35.3		

* See text for details of significance tests. Much of the remaining Atlantic and Pacific SST variability appears to be noise when viewed on the ocean basin scale.

Standardised rainfall time series

We have used three different standardised rainfall series in an attempt to represent seasonal rainfall in part or all of the northern Nordeste. Series A is composed of March and April rainfall totals derived from thirty stations (Hastenrath and Heller 1977), recently extended to cover the years 1913–85 (see Figure 21.1 for location of stations). Series B is composed of March to May rainfall totals created from data for Fortaleza (F) and Quixeramobim (Q). The two stations are in the far north of the region (see Figure 21.1). The chief advantage of this series is that it is the longest (1901–87) and the series can be updated very quickly since F and Q are main synoptic stations. Series C is composed of February to May rainfall totals for 114 stations compiled by Nobre and Paiao (INPE, Brazil). This series covers the years 1912–81. (It was not available when we released a forecast in March 1987.) Over the years where values for all three series are available (1913–81), the inter-correlation is moderately good, in the range 0.78–0.83. There are no significant serial correlations or trends, i.e. inter-annual variability is dominant, in contrast to the postwar Sahel rainfall series (Folland, Parker and Palmer 1986).

Sea-surface temperature patterns and northern Nordeste rainfall

For every 5° latitude x 5° longitude area in the Atlantic for which SST data were available, time series of SSTAs averaged over March and April were correlated with each of the northern Nordeste rainfall series. Figure 21.3 shows the spatial pattern of correlations with Nobre–Paiao rainfall series during 1946–81. The pattern suggests that drought (flood) is associated with a dipole of positive (negative) SSTAs in the north tropical Atlantic and negative (positive) SSTAs in the south tropical Atlantic with variations near the Benguela current region west of South Africa and Angola being particularly important. This is consistent with previous work mentioned above. The existence of significant correlations with north extra-tropical Atlantic SST indicates that a direct link may exist between anomalies in the tropics and extra-tropics in the Atlantic sector.

Figure 21.4 shows very similar spatial correlations, also with the Nobre-Paiao rainfall series, for the earlier period 1913–45. The similar pattern of significant correlations in the tropical Atlantic confirms the stability of the SST–rainfall relationship there. Further north, correlations are less significant but remarkably similar, with alternating bands of positive–negative–positive correlations extending north-east from the Gulf of Mexico towards Europe.

Figure 21.5 shows the third covariance EOF (A3) of December–February Atlantic SSTAs for the period 1901–80. It explains 7 per cent of the variance of the above data and is similar to the spatial correlation patterns in Figures 21.2 and 21.3. The close resemblance extends over both the tropics and the northern extra-tropics. Figure 21.5 implies, therefore, that northern Nordeste seasonal rainfall is linked to a principal inter-annual variation of SSTAs. The time series of A3 formed using March–May SSTAs has highly significant correlations with all three rainfall series for 1913–81 ($r = 0.49, 0.56, 0.55$ for series A, B and C respectively, all significant at the 99.9% level when conservatively assuming 50 degrees of freedom). A similar significant pattern of SST–rainfall correlations emerges when the average of the preceding January and February SSTAs are correlated with all three rainfall series (not shown). The time series of A3 measured in January and

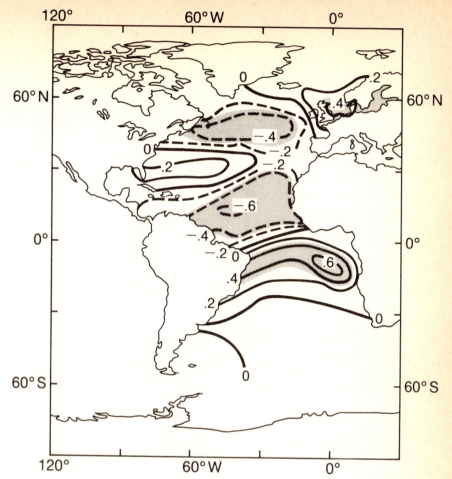

Figure 21.3 Correlation over 1946–81 between northern Nordeste rainfall (time series C) and the average of March–April SST Anomalies (5° × 5° grid square time series)

Note: Stippled areas are significant at 95%, assuming conservative estimate of 30 d.f.

February (plotted on Figure 21.6) is, not surprisingly, also well correlated with the subsequent seasonal northern Nordeste rainfall ($r = 0.49$–0.56 depending on rainfall series).

Spatial correlations were also calculated using Pacific SSTAs. The pattern of correlations (not shown) identified the El Niño event (see Gregory, Chapter 22, this volume) as a useful potential predictor of northern Nordeste drought. Cold eastern tropical Pacific events, with anomalies of opposite sign to El Niño events, were associated with wet conditions in northern Nordeste. The first Pacific EOF (P1) (Figure 21.7) reproduces these SSTA patterns well. The time series of P1 (Figure 21.8) traces the history of El Niño and cold events in the Pacific and, using January and February SSTAs, has correlations of 0.34, 0.28 and 0.33 with series A, B and C respectively between 1913–81 (all significant at the 95% level assuming 50 degrees of freedom).

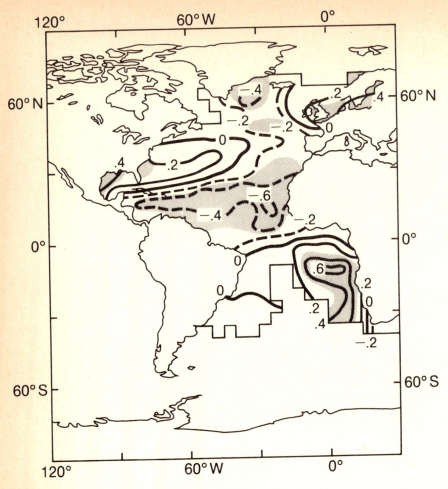

Figure 21.4 As Figure 12.3 but for 1913–45

Note: Areas with significant SST data for analysis are boxed.

The physical link between Pacific SSTs and Nordeste rainfall is likely to be complex and may arise through more than one physical effect. One aspect seems to be that the presence of El Niño/cold events in January is significantly related to subsequent changes in Atlantic SSTs in areas important for Nordeste rainfall (Figure 21.9). This is consistent with recent findings by Wolter (1987) who related the Southern Oscillation Index to tropical Atlantic SSTAs.

Forecasting northern Nordeste rainfall

We have attempted to forecast northern Nordeste rainfall by using time coefficients of the Atlantic and Pacific EOFs as predictors. Here we discuss models containing predictor time coefficients formed from SSTAs averaged over the preceding

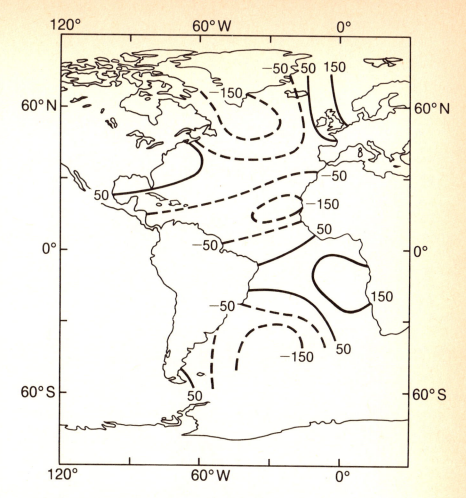

Figure 21.5 Third covariance eigenvector of Atlantic Ocean SSTA for seasonal December–February data, 1901–1980 (weights are × 10²)

Note: For a given point the eigenvector weights are such that when multiplied by the corresponding eigenvector time series coefficient and the products so calculated are summed over all eigenvectors the observed SSTA is recreated.

November to January or, separately, for February only (the EOF patterns themselves are unchanged). Stepwise linear discriminant analysis is used to establish predictive relationships between the rainfall series and SST EOF coefficients (predictors) over a long training period (for details of the method, see Maryon and Storey 1985). A particular year's November to January or February SSTA field is used to form predictor EOF coefficients which are used in the prediction equations to forecast northern Nordeste rainfall for the season immediately following. The forecast is currently expressed in terms of a set of posterior (forecast) probabilities for five rainfall categories. The categories, ranging from very dry to very wet, are chosen to be equiprobable in the training period (prior

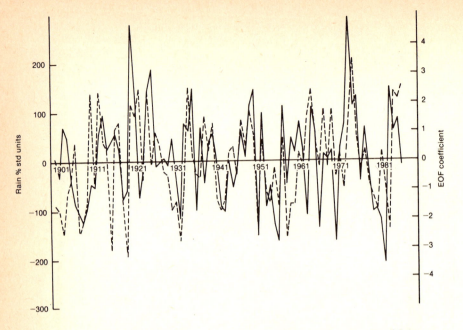

Figure 21.6 Atlantic EOF 3 (Figure 21.5) time coefficient using average of January and February SST data (solid line) along with Fortaleza–Quixeramobim rainfall series (dashed line)

probability of 0.2). Such categories can be referred to as 'quints' (Folland, Woodcock and Varah 1986).

Objective techniques have been used to help select that combination of predictor EOF time coefficients most likely to produce a prediction model that is most skilful in years outside the training period (for further details see Maryon and Storey 1985). A three-predictor model is chosen here, including EOFs A3 and P1 and, less importantly, an Atlantic EOF pattern (EOF A4) which has weights of similar sign across the tropical Atlantic suggesting an influence of the average temperatures of the whole tropical Atlantic (A3 and A4 are not completely statistically separated (Table 21.1), a problem needing further investigation).

Because rainfall and SST data are available for about the last 80 years, a forecast model constructed using a training period of about 40 years can be tested on a further 40 years of independent data, giving long testing and long training periods. This type of experiment has been performed using models based on a pre-1945 training period to make forecasts for years after 1945 and vice versa. Models have been created separately for Hastenrath's and the F and Q rainfall series.

Categorical forecast verification

The rainfall quint predicted to have highest probability in a given year is taken to be the 'best estimate' forecast. Table 21.2 summarises the average performance of four models assessed in this mode using November to January predictor patterns.

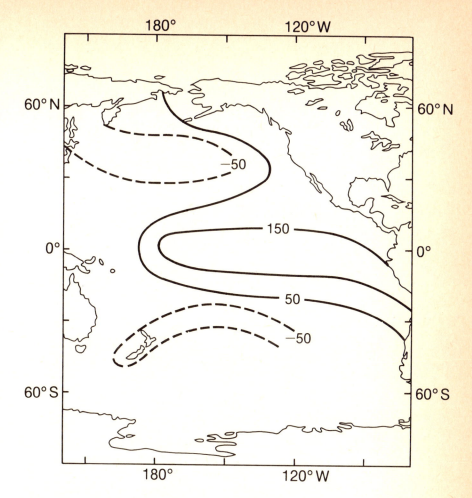

Figure 21.7   First covariance eigenvector of Pacific Ocean SSTA for seasonal
December–February data, 1901–80

When the models predicted a very dry year, northern Nordeste rainfall was
observed to be dry or very dry on almost 65 per cent of occasions, compared to
a chance level of 40 per cent. Table 21.3 shows that, using February SST patterns,
75 per cent of quint 1 forecasts were made for years subsequently observed to be
dry or very dry. Skill in predicting above average rainfall is also evident, but years
of near-normal rainfall are poorly predicted.

Table 21.4 shows the correlation between the best estimate forecast quint and the
observed quint for each forecast experiment in Tables 21.2 and 21.3. The correlations are encouraging, especially as the forecast is sometimes being made from
models established over a training period far removed from the year being forecast.
For example, one would not normally forecast the 1970s from a model established
only on 1901–45 data but would include more recent data in the training period.

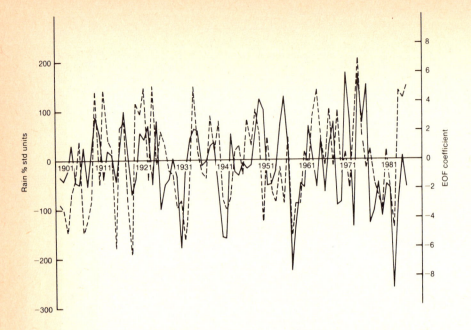

Figure 21.8 Pacific EOF one (Figure 21.7) time coefficient using the average of January and February SST data (solid line) along with Fortaleza–Quixeramobim rainfall series (dashed line)

Probability forecast verification

Viewing this category that is forecast with highest probability as a categorical prediction is not the only or most fundamental way of evaluating these forecasts. Instead, account may be taken of the forecast probability of each category.

The ranked probability score (RPS) makes such as assessment by summing over each quint the squared difference between the forecast cumulative probability and the observed cumulative probability. The latter is a step function rising from zero to one at the cumulative probability of the observed quint. The smaller the summation the better the forecast. Here, the RPS score for a set of forecasts is used in the normalised form discussed by Daan (1985) where the score is expressed relative to one achieved using a random forecast strategy; sums of the scores have then been converted into the percentage of that achieved by a set of perfect forecasts, i.e. those giving a forecast probability of one to the observed category. The RPS score is fairly harsh because a forecast probability of one is rarely even approached. Table 21.4 shows the scores of the forecast experiments in Tables 21.2 and 21.3. Scores approaching 30 per cent in fact indicate considerable skill. Prediction models for seasonal Sahel rainfall (Parker, Folland and Ward, Chapter 15, this volume) score quite similarly at about 25 per cent, while the recently issued UK monthly long-range probability forecasts of temperature and rainfall have an average score of only about 6 per cent on this searching measure (see Figure 21.10).

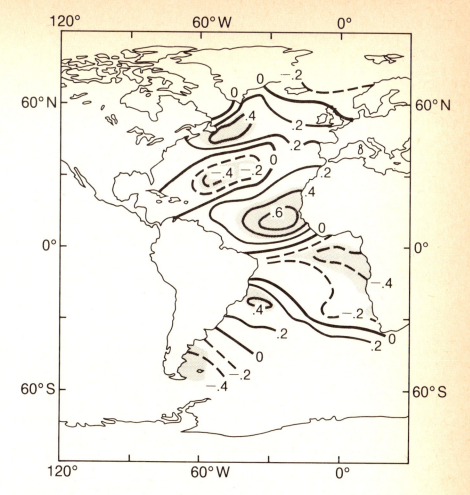

Figure 21.9 Correlation over 1946–86 between El Niño eigenvector coefficient in January and subsequent change in SSTA January to May (5° × 5° grid square time series)

Note: Stippled areas are significant at 95% level, assuming conservative estimate of 30 d.f.

Conclusions and experimental forecast for 1987

Work reported here and elsewhere in this volume offers a reasonable prospect of practically useful forecasts of seasonal rainfall in certain tropical regions. A forecast of a very dry (quint 1) northern Nordeste rainfall season was released in March 1987 (see Tables 21.5 and 21.6). Provisional rainfall totals for March, April and May at Quixeramobim and Fortaleza are shown in Table 21.7, along with 1951–80 normals. Despite a wet March, the three-month period is among the driest 25 per cent of rainfall seasons in the whole (1901–86) F and Q series and is just among the driest 20 per cent in the period 1946–86; it is therefore on the boundary between quints 1 and 2. The forecast for the F and Q rainfall series has therefore

Table 21.2 Summary of the average performance of northern Nordeste forecast experiments using November–January SSTS

		Observed quint				
		1	2	3	4	5
Forecast	1	17	8	6	5	3
quint	2	4	7	8	6	3
	3	7	7	7	6	4
	4	5	6	9	3	11
	5	0	1	4	12	11

List of experiments:
 (i) 1913–45 training data, forecasts for 1946–86, Hastenrath's March–April rainfall series
 (ii) 1946–86 training data, forecasts for 1913–45, Hastenrath's March–April rainfall series
 (iii) 1901–45 training data, forecasts for 1946–86, Fortaleza/Quixeramobim March–May rainfall series
 (iv) 1946–86 training data, forecasts for 1901–45, Fortaleza/Quixeramobim March–May rainfall series
The predictors are the 1901–80 ocean basin winter SST anomaly EOF coefficients representing the observed strength of Atlantic EOF patterns 3 (Figure 21.5) and 4 and Pacific EOF pattern 1 (El Niño) in the November–January SSTA field preceding the rainfall season.

Table 21.3 Summary of the average performance of northern Nordeste forecast experiments using February SSTS

		Observed quint				
		1	2	3	4	5
Forecast	1	22	8	6	3	1
quint	2	4	11	6	3	4
	3	6	2	6	5	6
	4	1	3	8	12	12
	5	0	5	8	9	9

Table 21.4 Skill scores for a set of forecast experiments*

Rainfall series	Training period	Test period	SSTs	RPS skill (%)	Forecast–Observed Quint Correlation
Hast.	1913–45	1946–86	Nov.–Feb.	28.5	0.44*
			Feb.	27.0	0.62[†]
	1946–86	1913–45	Nov.–Feb.	25.9	0.53[†]
			Feb.	29.2	0.53[†]
F/Q	1901–45	1946–86	Nov.–Feb.	26.5	0.54[†]
			Feb	20.0	0.58[‡]
	1946–86	1901–45	Nov.–Jan.	21.3	0.47[†]
			Feb.	23.4	0.34

 * Correlation significance assumes 30 d.f.
** At 95% level
 [†] At 99% level
 [‡] At 99.9% level

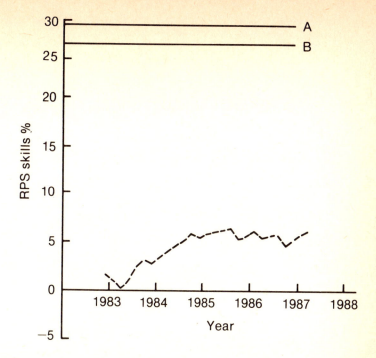

Figure 21.10 Graph illustrating the order of magnitude difference in skill between long-range forecasts for northern Nordeste and UK using the ranked probability score

Note: Dashed line shows four-year running mean for UK monthly precipitation forecasts. Solid line show mean score of northern Nordeste seasonal rainfall predictions using February SSTs. A: forecasts for 1946–86, B: forecasts for 1913–45 (see Table 21.4).

Table 21.5   Forecasts of northern Nordeste rainfall (Hastenrath's March–April series) in 1987*

SSTA field used to make the forecast	Forecast probability for 1987 rainfall season				
	Quint 1 Very dry	Quint 2 Dry	Quint 3 Average	Quint 4 Wet	Quint 5 Very wet
Nov. 1986–Jan. 1987	0.57	0.19	0.13	0.07	0.04
Feb. 1987	0.61	0.18	0.15	0.03	0.02

*Probability distributions shown are the average of five forecasts based on training periods 1913–45, 1913–82, 1913–86, 1946–82, 1946–86. The quint boundaries were defined from the training periods concerned.

Table 21.6   Forecasts for Quixeramobim and Fortaleza rainfall (standardised March–May total) in 1987*

SSTA field used to make the forecast	Estimated probability for 1987 rainfall season				
	Quint 1 Very dry	Quint 2 Dry	Quint 3 Average	Quint 4 Wet	Quint 5 Very wet
Nov. 1986–Jan. 1987	0.45	0.13	0.18	0.23	0.01
Feb. 1987	0.45	0.18	0.14	0.18	0.05

*Probability distributions shown are the average of five forecasts based on training periods 1901–45, 1901–82, 1901–86, 1946–82, 1946–86.

Table 21.7   Initial monthly rainfall totals (mm) for Quixeramobim and Fortaleza during the 1987 rainfall season

Month	Quixeramobim		Fortaleza	
	1987 Observed	1951–80 Normal	1987 Observed	1951–80 Normal
March	235	166	414	305
April	61	160	197	261
May	19	115	55	207
Total % of normal	71		86	

Source: Meteorological Office of Ceara State, Brazil

been substantially correct. Concentration of the rainfall deficit into the later part of the rainfall season appears to have had a particularly severe impact on agriculture in the region. Current techniques are, of course, not designed to predict the intraseasonal rainfall variations which are clearly very important.

Verification of the forecast based on Hastenrath's 30-station northern Nordeste rainfall season awaits receipt of the data. Current work is addressing the way in which El Niño events affect the SST in the tropical Atlantic during the Nordeste rainfall season (see also Wolter 1987). Refinements to the eigenvectors are also in hand.

References

Covey, D.L., Hastenrath, S. 1978. 'The Pacific El Niño phenomenon and the Atlantic circulation', *Mon. Wea. Rev.* 106, 1280–7.

Craddock, J.M., Flood, C.R. 1969. 'Eigenvectors for representing the 500 mb geopotential surface over the northern hemisphere'. *Q.J.R. Met. Soc.* 95, 576–93.

Daan, H. 1985. 'Sensitivity of verification of scores to the classification of the predictand'. *Mon. Wea. Rev.* 113, 1384–92.

Folland, C.K., Parker, D.E., Palmer, T.N. 1986. 'Sahel rainfall and worldwide sea temperatures 1901–85', *Nature* 320, 602–7.

Folland, C.K., Woodcock, A., Varah, L.D. 1986. 'Experimental monthly long-range forecasts for the United Kingdom. Part III Skill of the monthly forecasts'. *Met. Mag.* 115, 377–95.

Hastenrath, S., Heller, L. 1977. 'Dynamics of climatic hazards in Northeast Brazil', *Q.J.R. Met. Soc.* 103, 77–92.

Hastenrath, S., Wu, M.-C., Chu, P.-S. 1984. 'Towards the monitoring and prediction of Northeast Brazil droughts'. *Q.J.R. Met. Soc.* 110, 411–25.

Lau, N.C., Oort, A.H. 1985. 'Response of a GFDL general circulation model to SST fluctuations observed in the Tropical Pacific Ocean during period 1962–1976' in Nihoul, J.C.J. (ed.) *Coupled Ocean-Atmosphere Models*, Amsterdam, Elsevier, 289–302.

Maryon, R.H., Storey, A.M. 1985. 'A multivariate statistical model for forecasting anomalies of half-monthly mean surface pressure', *J. Climatol.* 5, 561–78.

Moura, A.D., Shukla, J. 1981. 'On the dynamics of drought in Northeast Brazil. Observations, theory and numerical experiments with a general circulation model', *J. Atmos. Sci.* 38, 2653–75.

Namias, J. 1972. 'Influence of northern hemisphere general circulation on drought in Northeast Brazil', *Tellus* 24, 336–43.

North, G.R., Bell, T.L., Cahalan, R.F., Moeng, F.J. 1982. 'Sampling errors in the estimation of Empirical Orthogonal Functions', *Mon. Wea. Rev.* 110, 699–706.

Palmer, T.N. 1986. 'Influence of the Atlantic, Pacific and Indian Oceans on Sahel rainfall', *Nature* 322, 251–3.

Wolter, K. 1987. 'The Southern Oscillation in surface circulation and climate over the Tropical Atlantic, Eastern Pacific and Indian Oceans as captured by cluster analysis', *J. Clim. Appl. Met.* 26, 540–58.

Chapter 22

The effects of ENSO events in some key regions of the South American continent

N. Caviedes   University of Florida, USA

Today the climatic variations that have come to be known as El Niño/Southern Oscillation events do not need an introduction — it is well known that the anomalies that arise in the tropical Pacific during particular years have far-reaching implications and that they are among the major climatic variations of our times.

It remains to be seen — and studies on the matter are in progress — in which way these anomalies are reflected in the pressure and rainfall patterns of various regions of South America, and at what time they appear in relation to the inception of ENSO events in the tropical Pacific — the first zone to show signs of upset circulation conditions. In this chapter an examination — limited by the space available — will be made on the progression of such anomalies in some key areas of South America.

An initial assumption is that the South Pacific anticyclone (SPA) and its associated wind systems dominate most of the oceanic-atmospheric circulation of western South America. This centre of action is not only crucial for the existence or absence of coastal upwelling and the strength of the Peru Current, but it is also responsible for the trade winds in the south-eastern Pacific. Fluctuations in the strength of the South Pacific anticyclone (Caviedes 1985), measured by conventional tools, such as the 'Southern Oscillation Indices', reveal better than any other measure the inception, maturity and decline of ENSO events (Wright 1984). It must be stressed then, that barometric considerations, which indicate the dominance of the atmospheric forcing over oceanic circulation and sea-surface temperatures (and, indirectly, over precipitation), are the basis for any distinction between normal and abnormal conditions in the tropical Pacific and South America.

Moreover, with ENSO events being primarily anomalies of the tropical circulation, their occurrence can be expected to be the first reflected by the pressures and rainfall in the South American tropics and subtropics. With this in mind, we selected a group of stations in six major 'natural-geographic' regions of South America (Figure 22.1) whose rainfall regimes have been adequately documented (Caviedes 1981).

1. Oceanic subtropical South America is represented by the achipelago of Juan Fernandez, at 36 °S, off the coast of central Chile. Easter Island (27 °S) and Juan Fernandez are virtually the only stations that can gauge the barometric variations of the SPA, near its centre and at its south-eastern margin, respectively. In fact, both stations register the highest pressure values in the tropical and subtropical Pacific and its rimlands. Their maxima during the winter triad (JAS) and their minima in the southern summer (DJF) reflect a well-defined yearly cycle which is altered only during anomalies.

2. Temperate central Chile, represented by Valparaiso, is a subtropical region directly influenced by the winds that emanate from the SPA and the cold

Figure 22.1   Stations representative of six natural-geographical regions in South America considered in this study

upwelling waters of the Peru Current. Some continental influences, particularly in summer, tend to alter somewhat the otherwise strongly maritime character of the climate of coastal central Chile. The highest pressure values occur at the end of summer, in winter and in early spring when frontal passages and cyclonic depressions decrease in frequency (Romero 1986). These winter occurrences — all of them generated in the subtropical Pacific — account for the winter rains that peak in the JJA triad. However, there are years when the rainy period begins in April (early autumn) and continues into September (early spring).

3. Northern Peru and southern Ecuador, represented by Piura, is the arid coastal region located at the northernmost extremity of the Peru Current, shortly before it turns away from the continent and becomes the Southern Equatorial Current. Under the blocking conditions determined by the air subsidence over the upwelling centres that feed the Peru Current and the drying effects of south-easterly winds, this region is an equatorial desert that shares many of the climatic characteristics of dry Pacific islands on the equatorial line, such as the Galapagos, Christmas, Malden or Canton Islands. In this domain, rains occur during the summer months (JFM), when the slackening of the south-eastern trades and the seasonal oceanic warming allow the episodic shift of the Inter-tropical Convergence Zone south of the equator. Yet, the inter-annual variability of precipitation ranges from years with scarce summer rains to years with copious and torrential rains: El Niño years (Mugica 1984). Indeed, this is the region of South America that bears the brunt of the oceanic-atmospheric anomalies now known as 'ENSO events'. It is the classic 'El Niño region' of South America, the one whose characteristics reflect best the atmospheric and oceanic anomalies of the eastern tropical Pacific.

4. La Paz, located in the high plateau of Bolivia and southern Peru, known as the Altiplano, is another key region in terms of regional anomalies of pressure and precipitation in South America. Isolated from the climatic controls of the Pacific ocean by a towering cordillera range and partially cut off from the humid air masses originating from the tropical lowlands of the continent by another range, the Altiplano receives most of its precipitation from upland convection during early summer (DJF). Although the inter-annual precipitation variability in the Altiplano is not as large as in northern Peru, there are years when the winter dryness extends into spring and summer, thus producing droughts (Francou and Pizarro 1985). It has been demonstrated that these droughts are especially pronounced during the years when northern Peru is struck by ENSO episodes (Caviedes 1982a), a fact that points to a potential linkage between these droughts and the positive precipitation anomalies of the classic El Niño region.

5. The Amazon basin, represented by Manaus, encompasses the vast tropical lowlands of the continental core. But since this is too gross a generalisation of the rainfall regimes and wind systems across the whole basin, a more pertinent regional distinction is made between an inner Amazon basin (with Iquitos and Uaupes as typical stations) which is submitted to the same mechanisms as the eastern slopes of the tropical Andes (Caviedes 1981), and an outer Amazon basin (with Manaus as its representative station) that reacts to the zenithal of the centre of the continent and that reflects some distant stimuli from the circulation and humidity budget of air masses generated in the tropical Atlantic (Vulquin 1971; Virji 1981). The stations located along the larger axis of the Amazon basin follow almost the equatorial line and should, therefore, reveal the circulation anomalies that take place in a zonal rather than a meridional direction.

6. North-eastern Brazil, another classic climatic region of South America, is represented by Quixeramobim, a station with long records and considered as typical of the arid north-east of Brazil (Markham and McLain 1977). Rainfall occurs during the peak of the southern summer (FMA), but when rains are scant or fail to occur, the region is beset by devastating droughts (*secas*) that upset not only a fragile natural environment but also the structures of society (Brooks 1981). The *secas* has been studied intensively (see the contribution of Ward, Brooks and Folland in Chapter 21 this volume) and have been tied to circulation anomalies over equatorial eastern South America (Nobre *et al.* 1985) and to oceanic anomalies in the tropical Atlantic (Chu and Hastenrath 1981). A remarkable coincidence in the years in which north-eastern Brazil is beset by droughts and the eastern tropical Pacific is affected by ENSO events has been long recognised (Doberitz 1969; Caviedes 1973) and has prompted several studies to document the existence of genetic linkages between these anomalies at the two extremes of equatorial South America.

Thus, these stations along the equatorial and subequatorial belt (0–15 °S), plus two stations from the oceanic extra-tropics of western South America, have been selected to offer a general insight into the manner in which circulation anomalies that develop in the tropical south-eastern Pacific expand and influence not only the rainfall mechanisms of Pacific stations but also the observed circulation and rainfall fluctuations in some significant natural regions of South America.

The methodology. Using crossing theory

The determination of El Niño years, normal years and anti-El Niño years (when the oceanic/atmospheric conditions are in extreme opposition to ENSO years, i.e. strong SPA, steady trade winds, cold ocean and lack of precipitation) has been accomplished not without some disagreement among climatologists, oceanographers and ecologists. Not every year that exhibits oceanic warming in the tropical Pacific can be qualified as an 'El Niño' year even if summer rains fail to occur in northern Peru or the Peru Current and its related upwelling centres are strong. On the other hand, there have been some years during which it has rained abundantly in southern Ecuador and northern Peru, in circumstances that no trace of warm water was found in the eastern tropical Pacific (Waylen and Caviedes 1986).

An examination of the oceanic conditions in the eastern tropical Pacific and of rain amounts and river discharges in northern Peru for a series of 75 years (1911–85) enables us to pinpoint years in which the atmospheric circulation and the oceanic conditions of the eastern tropical Pacific and of western South America were noticeably upset (ENSO events), years in which ocean and atmosphere did not display remarkable variations from the acceptable norms (normal years) and years in which the ocean was abnormally cool, the Peru current particularly strong and the summer rains in northern Peru scarce or non-existent (anti-ENSO years). From these years (Table 22.1), we selected for close scrutiny the most recent 'major ENSO' occurrences based on the fact these are the most accurately recorded: these were the ENSO events of 1972–3 and 1982–3, and the controversial ENSO episode of 1977–8 was added to provide a means of comparison between major and minor ENSO events.

Table 22.1 Years of occurrence of ENSO events, normal ocean-atmospheric conditions, and anti-ENCO conditions

ENSO events	Normal	years	Anti-ENSO	years
1911	1915	1940	1912	1966
1925	1916	1942	1913	1967
1926	1917	1944	1914	1968
1932	1918	1945	1921	1970
1933	1919	1946	1924	1974
		1948	1927	
1939	1920	1949	1930	1979
1941	1922	1952	1937	1980
1943	1923	1955	1947	1984
1953		1956		
1957	1928	1959	1950	1985
1965	1929	1969	1951	
1972	1931	1971	1954	
1973	1934	1975	1958	
1977	1935	1976	1960	
			1961	
			1962	
1978	1936	1981	1963	
1982	1938		1964	
1983				

Source: Adapted from Waylen and Caviedes (1986). The years of occurrence of major ENSO events are printed in italics.

Since the main concern of this investigation is to pinpoint the time when variations in the normal patterns of pressures can be considered as the inception of an anomaly and to determine the intensity and length of such fluctuations, 'crossing theory' — an exploratory tool successfully utilised in hydrology to discern extreme flood events from high discharge in rivers — was chosen to perform the analysis (Todorovic 1978; Waylen 1985).

In this study 'crossing theory' is applied to a time series of surface-level pressures in order to determine the 'abnormal pressures', i.e. those that are above or below a selected *truncation level* Q_o. Customarily the truncation level adopted is a certain number or fraction of the standard deviation (s). Each abnormal pressure occurrence is characterised by a 'peak magnitude' (H_1) which is the number of millibars above the truncation level reached by extreme events, like Q_{p1} or Q_{p2} (Figure 22.2).

The temporal distance between the beginning of an 'upcrossing' and its return below the truncation level m_1 or m_2 is the duration of the anomaly. Upcrossings and downcrossings constitute a *partial duration* series consisting of two random variables: the number of crossings per year, and the peak magnitude of each event.

As to the magnitude of events, the exceedance magnitudes of pressure series can be fitted to a theoretical exponential distribution, derived from the mean exceedance magnitudes of upcrossings and downcrossings. This allows the qualification of the character and distribution, in millibars, of extreme events.

However, in this chapter emphasis will be placed on the historical sequence of pressure variations as a means for detecting the beginning and end of the ENSO events of 1972–3, 1977–8 and 1982–3 at selected South American stations. Since drastic rainfall variations measured in monthly intervals reveal clearly the impact

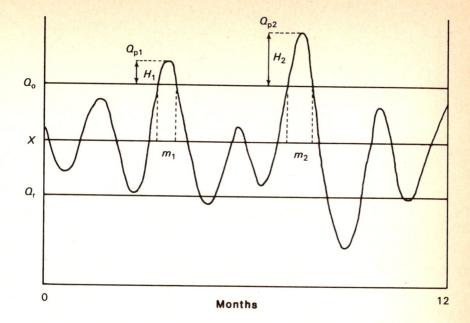

Figure 22.2   Variables used in partial duration series analysis

X = mean value
Q_o = upper truncation level
Q_r = lower truncation level
The symbols corresponding to other variables are explained in the text.

of ENSO events, monthly precipitation totals have been included in the diagrams illustrating the pressure fluctuation, but their crossing analysis is not performed here. The beginning of each ENSO event has been set in July of the year preceding an ENSO event, for our preliminary analyses of the time series tend to suggest that the first signs of upset barometric and circulation conditions appear in the southern winter; this norm applies particularly to the pressures series of Juan Fernandez and Valparaiso. The series considered are continued until the end of the autumn (May and June) that follows the peak of the event. In this manner, there are two 'southern summers' included, a fact which is convenient for the analysis of ENSO episodes since it allows the evolution of such events to be followed into their maturity (December through April), and to establish the features of their autumnal decline. If such decline fails to occur, the anomalies continue into the next southern summer.

El Niño 1972–3

In Pacific stations (Juan Fernandez and Valparaiso) the prelude of El Niño was marked by abnormally high pressures at the peak of the southern winter (notice the upcrossings in JAS), Figure 22.3. This was followed by a brisk descending tendency that culminates in NDJ of 1971–2. At the height of El Niño 1972 (JFM),

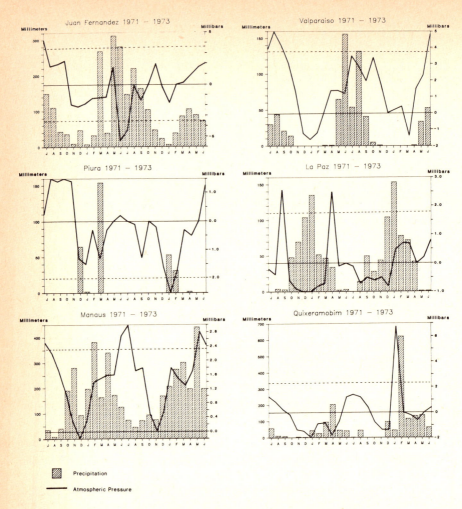

Figure 22.3   Pressure and rainfall series of the ENSO event 1972–3*

* The series begins on July 1971.

Valparaiso continued to show low pressure values — though no downcrossings — whereas Juan Fernandez maintained relatively low barometric levels that indicated a weakened South Pacific anticyclone during the summer of that year. Piura exhibited high pressure values during the spring (SON) preceding El Niño and then an abrupt descent — also observed in Valparaiso — during December 1971. In that month, and also in March 1982, conspicuous pressure declines and concentrated precipitation suggest the influence of a southward shifted ITCZ or the continuous onslaught of convecting systems that have glided on top of warm waters into the domain otherwise dominated by the cold Peru Current. That the autumn and winter of 1972 did not bring the seasonal recovery of the SPA and its associated wind systems is observable in Juan Fernandez in another 'collapse' of the winter pressure levels, which even attained a downcrossing during June–July 1972, and in Piura where the pressures did not rise above the average winter levels. Valparaiso presented the characteristic ups and downs of a weak SPA, and the unusually early and copious winter rains of 1972 are to be considered as the last consequences of an abnormally warmed south-eastern Pacific during the previous season. In fact, contrary to El Niño 1982–3 in many Pacific stations, El Niño 1972–3 was more evident in oceanic and ecological disturbances than in dramatic rainfall increases.

The Altiplano (La Paz) exhibited a pressure rise during August–October 1971, and even an upcrossing in September, very much in tune with the thesis that during ENSO episodes in northern Peru high pressure levels and lowered precipitation are to be expected in the Altiplano. However, the latter did not hold true, for it rained abundantly during the summer of 1972–3. But from February to April there occurred a drop in precipitation and a marked increase in pressures which may indicate a minor drought on the Andean plateau. During the winter and spring of 1982, the pressures continued to be low and the onset of the next rainy season at a relatively early date (August–September) heralded a return to normal climatic conditions over the Altiplano.

In the interior of the Amazon, barometric pressure was rising during the southern winter (note the upcrossings in July) and followed by a sharp decline during spring which was coupled with a decrease in precipitation amounts. A low pressure level in December is very suggestive of a possible link between the low pressure trends that were affecting the equatorial eastern Pacific, southern Ecuador and northern Peru during December 1972. The following obvious upward jump in February — which was in phase with the pressure drops experienced by the stations of northern Peru — revealed even more clearly the 'see-saw' relationship that theoretically exists between excessive rain in coastal Ecuador and Peru and lowered precipitation in the centre of the Amazon basin (Caviedes 1982a). With the arrival of the southern winter the pressures soared to an upcrossing in June–July so conspicuously as to be considered a compensation for the lowered barometric levels that were experienced on the equatorial coast of the Pacific, on the other side of the Andes.

In Quixeramobim, north-eastern Brazil, there were no indications of an increase in pressures from the winter of 1971 into the early summer of 1972 as the theory would demand. But the extremely scarce precipitation that fell during the summer months of 1972, especially during the JFM triad, are unmistakable signs of the inverse relationships that exist between ENSO episodes on the Pacific coast of equatorial South America and the droughts of north-eastern Brazil.

After March–April 1973, all the Pacific stations recorded upward tendencies in pressure that signalled the recovery of the main centre of action in the south-eastern Pacific and the restoration of the normal air-ocean circulation patterns. The powerful upcrossing of Valparaiso and the upward trend of Piura and Juan Fernandez in June 1973 marked the onset of a 'pressure rebound' that was to culminate in the anti-El Niño episode of 1974.

El Niño 1977–8

In Pacific stations, such as Juan Fernandez, the upset barometric conditions in the south-eastern Pacific can be documented by the drastic drop and downcrossing in September 1976 (Figure 22.4). After rather strong anticyclonic conditions during the winter of 1976, Valparaiso showed a similar falling tendency from October

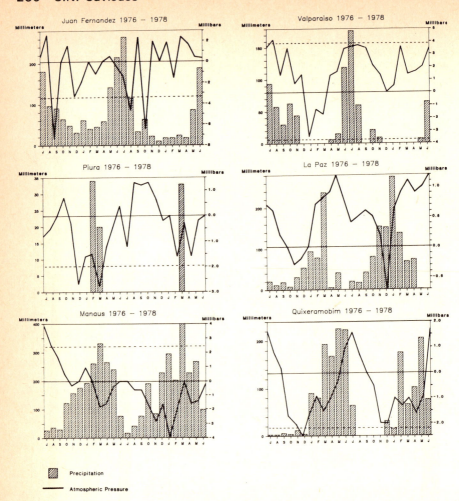

Figure 22.4 Pressure and rainfall series of the 'aborted' ENSO event 1977–8*

* The series begins on July 1976.

1976 to January 1977, when a downcrossing almost did occur. These were clear symptoms of the early fluctuations of a weakened South Pacific anticyclone. The downcrossings of north Peruvian stations during December 1976 and March 1977 seemed to indicate that an ENSO anomaly was in the making, and the expectation was strengthened by the occurrences of copious summer rains in Piura during February and March 1977. However, after these summer events, and when the warming of the equatorial Pacific failed to reach the coast of Ecuador and Peru and remained confined to a band west of the Galapagos Islands (Wyrtki 1979), it became patent that the heralded ENSO episode had evolved into an 'aborted El Niño'.

That a short-lived weakening of the SPA and an areally reduced tropical Pacific warming had taken place was also illustrated by the intense rains in Juan Fernandez and Valparaiso during June and July 1977 and by a renewed up-and-down trend in the pressures of Juan Fernandez from June through November of the same year. In Piura the winter and spring recovery was very evident, and although in the ensuing summer there was a slight drop in pressures and some rain in March 1978, these variations can hardly be ascribed to a major ENSO episode.

In the Altiplano, La Paz showed falling pressures during the second half of 1976 and reduced summer precipitation during the DJF triad, but, after this interlude, the pressures soared and it rained too much in March for that summer to be qualified as a 'dry summer'. Moreover, the barometric levels of the winter of 1977 and the scarcity of rains during that season indicate a restoration of seasonally normal conditions in the highlands of southern Peru and Bolivia. In the following summer, in spite of a dramatic drop in pressures near the lower truncation level in December, precipitation was too high to qualify that summer in the Altiplano as a dry season.

In the Amazon basin the pressures were high during the preparation phase. Then, there was a falling tendency that began in ASO 1976 and culminated in MAM 1977 with high precipitation amounts. During autumn 1977 and early summer 1978, pressure levels dropped remarkably and the rains were so copious that one can hardly recognise the negative precipitation variations that characterise years with ENSO episodes in the tropical Pacific.

North-eastern Brazil showed a falling barometric tendency all through the second half of 1976 and reached its nadir in December with a downcrossing. Very low levels of precipitation seemed to announce a very dry summer season, but when the pressures began to rise in DJF and the rains set in, there was no element that was a response to the aborted ENSO in the south-eastern pacific. During the winter there was the normal seasonal rise of pressures and absence of precipitation, followed by a new barometric drop in the spring of 1977. However, when the rains returned in March 1978, there was no doubt that that summer did not bring another dreaded drought either.

Towards the end of 1977, the SPA fluctuations (as illustrated by the pressures of Juan Fernandez) could have been interpreted as the delayed signs of an ENSO in the south-eastern Pacific, but the normalisation of barometric conditions became very evident during the first half of 1978. The rest of the stations also reflected a normality pattern with stabilised pressures and rainfall returning to seasonal norms. In sum, it can be said that — with the exception of the pressure drops and rainfall increases in northern Peru during the summer of 1977 — the regions of South America did not exhibit clear-cut variations so as to qualify that year as a major ENSO year.

El Niño 1982–3

In the subtropical Pacific stations (Juan Fernandez, Valparaiso and Piura) the end of winter and spring 1981 were characterised by extremely high pressures, and even a full OND triad with upcrossings in Valparaiso. All this indicated a strengthened condition of the SPA. Nevertheless, in January–February 1982, a drastic drop in pressures seemed to herald a weakening of the anticyclone. In fact, while in Valparaiso and Piura there were some signs of recovery in the following autumn and winter, Juan Fernandez experienced upset conditions during a long period of downcrossing that spanned from June to October 1982 (Figure 22.5). At

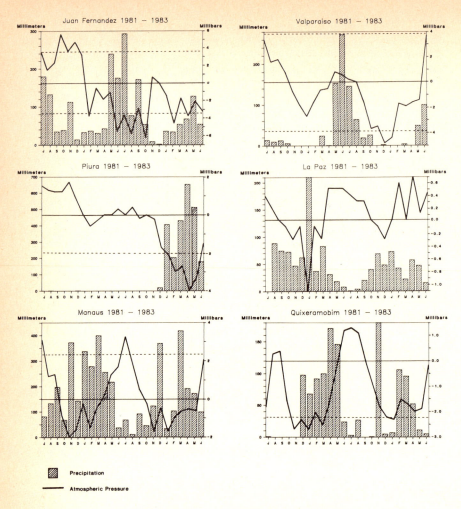

Figure 22.5   Pressure and rainfall series of the ENSO event 1982–3*

* The series begins in July 1981.

that station, as well as in Valparaiso, the weakening of the SPA was coupled with abnormally high winter precipitation.

A downward trend for all stations started in Valparaiso in August, and in Piura in October 1982. This was, precisely, the period in which the progressive warming of the tropical Pacific began to be monitored (Rasmussen and Hall 1983; Cane 1983).

Valparaiso registered downcrossing pressure levels in December 1982–January 1983, while in the north Peruvian stations, exemplified by Piura, the downcrossing period lasted until June 1983. The severe pressure drops were accompanied by the highest rains ever recorded instrumentally in southern Ecuador and northern Peru, and by an unparalleled oceanic warming of the south-eastern Pacific (Halpern and Hayes 1983).

In La Paz it is interesting to note that, after the normal winter of 1981, a sharp drop in pressures took place which culminated in January 1982, i.e. in phase with the strong anomalies that were affecting the south-eastern Pacific and its South America rimlands. After this drop, which was accompanied by a rather rainy January, there was a clear tendency towards recovery — with above average pressures — during the months when El Niño reached its maturity in the eastern Pacific: April–June 1982. The winter of that year was characterised in La Paz by relatively high pressures, followed by a gentle drop during October–December, and a powerful recovery during summer 1983 that was accompanied by abnormally low precipitation amounts. In the light of these features it can be generalised that in the Altiplano the echoes of an extremely upset meteorological condition in the south-eastern Pacific were felt during the JFM triad of 1982, after which a remarkable rebound to high barometric levels took place. The latter tendency is, in fact, more in tune with the high pressure levels and decreased rainfall that might be expected in the Altiplano during years with ENSO anomalies.

In the winter of 1981, the Amazon basin, as exemplified by Manaus, also recorded the extremely high pressures that were observed in stations of the subtropical Pacific (Juan Fernandez and Valparaiso). Nevertheless, during the spring (SON), a sharp decrease in pressures again reflected the anomalies in progress in the equatorial eastern Pacific. Rainfall levels were unusually high, refuting the assumption that ENSO events should be coupled with lowered precipitation and relatively high seasonal pressures in the Amazon basin. However, the rebounding that took place after the El Niño peak and culminated with an upcrossing at the end of the 1982 winter (August) is interesting. This trend was also observable during the ENSO event of 1972–3 — but not so during the mild ENSO episode of 1977–8 — so that it appears that, during the early stages of an ENSO in the eastern Pacific, the anomalies affecting the western margins of the Andes are so powerful that they influence the centre of the Amazon basin itself, with the elevation of the Andes in no way representing a barrier to high-level perturbations. But, once the abnormally high pressure levels are established in the Amazon, during the period following an ENSO event, their consequences are felt in a sizeable decrease of the spring and summer precipitation: notice the relatively scarce rains from August 1982 to June 1983 (December 1982 excepted).

In north-eastern Brazil, after a period of relatively high pressure levels during the winter of 1981 (JAS), a conspicuous drop below truncation levels took place from November 1981 to March 1982, i.e. in the preparatory ENSO phase. Fairly high summer precipitation occurred along with unusually low barometric levels. The core of the 1982 winter (JJA) brought a strengthening of pressure conditions, and despite a new falling tendency during the following spring (SON) and early summer (DJF), the extremely low rainfall of the summer 1982–3 (November excepted) left no doubts that El Niño in the eastern Pacific was, indeed, phased with a drought in the Brazilian Northeast. A feature which was not consistent with previous assumptions is that the critically low summer precipitation was coupled with relatively low barometric levels.

Unlike other ENSO events that dwindled in April or at the latest in May, the 1982–3 ENSO continued well into the winter of 1983. Juan Fernandez still exhibited downcrossings in June and July 1983, and in northern Peru it rained intensively in June. Only Valparaiso — among the Pacific stations — showed the typical 'rebound' to strong anticyclonic conditions in the winter of 1983 (June). In fact, along coastal stations of South America the oceanic warming lasted into

August–September of 1983, thus revealing the persistence of the ocean anomalies.

The Amazon stations also recorded a brisk return to normal pressure conditions in June 1983, very much in phase with the Pacific stations. The recovery of north-east Brazil occurred somewhat later, in August 1983.

The traits of inception and decline of ENSO events in South America

On the basis of the evidence obtained from the ENSO occurrences of 1972–3 and 1982–3, some generalities can be outlined. The islands located off the South American west coast, such as Easter Island or Juan Fernandez, record the inception of anomalies by means of brisk barometric pulsations during the autumn and winter preceding an ENSO event. Prior to these occurrences the Pacific Stations show extremely high pressure values that tend to suggest a strengthened South Pacific anticyclone and powerful south-easterly winds, very much in accordance with the opinions of Wyrtki that these conditions concur to produce the build-up of water in the western Pacific which is advected eastwards, later on, when the SPA weakens and the trade winds slacken (Wyrtki 1975; Lukas et al. 1984).

During the autumn and spring immediately preceding an ENSO event, pressures drop below the lower truncation levels and copious rains occur in the temperate regions of the continent. This pattern was particularly visible in 1982 when Juan Fernandez recorded pressure below the lower truncation level from June to October, heralding the anticyclonic weakening of the whole south-eastern Pacific. It can be hypothesised that the weak SPA permits frequent invasions of subtropical depressions in a region otherwise dominated by anticyclonic stability and cold waters, and that the polar front is able to advance farther north along the west coast of temperate South America because of the absence of anticyclonic blocking.

After a relative recovery of the pressures in the following spring, repeated drops and increased rainfall signal the expansion of the anomaly to tropical western South America: Piura exhibited downcrossings in January and February of 1973, and a very pronounced period below the lower truncation level from January to May 1983 as a result of the expanding El Niño phenomenon. In the centre of the Amazon basin the lowering of pressures occurs between October and December, almost as a continuation of the equinoxial lows of September: however, dryness is associated with these lowerings.

During the ensuing summer and early autumn generalised low pressures and torrential rains reveal the culmination of the event in northern Peru and southern Ecuador and its highlands.

Meanwhile, the outer Amazon basin, the north-east of Brazil and the high plateaux of Bolivia and Peru are beset by droughts that occur under relatively high pressure conditions. During extreme ENSO events, low pressure and decreased precipitation seem to be the norm; but these periods are immediately followed by energetic rebounds to high pressures. The phasing of El Niño droughts in north-eastern Brazil and the Altiplano thereby acquires a real significance.

During the winter and spring following a major ENSO episode conditions return to 'normal' in a brisk manner. It is, therefore, not uncommon that years with extreme ENSO events in the eastern Pacific have been followed by anti-El Niño events, if not in the same year, then the year after. Not to be forgotten in this context is the inertia of oceanic systems. The by no means fortuitous and drastic changes in pressures and ocean conditions that develop before and after an ENSO episode have prompted the proposition that all the dynamic, radiative and air-ocean circulation elements that interact to produce an ENSO event appear to bring about changes in the natural equilibrium that can, at least, be descriptively explained by a 'cusp catastrophe' model (Caviedes 1982b). The proofs of such a proposition depend very much on the parametrisation of all the elements involved in the genesis of ENSO phenomena.

If it is accepted that most major ENSO events show their first signs at the end of the southern winter or in early spring, then it can be posited that the cycle of inception, maturity and decline of ENSO events tends to last between 14 and 20 months, a time-span that may include two consecutive southern summers. Considering this time expanse, it is understandable that major ENSO events that cause two summers of torrential rains or two summers of dryness in other regions constitute the most feared climatic anomalies on the South American continent.

References

Brooks, R. 1981. 'Drought and adjustment dynamics in north-eastern Brazil', *GeoJournal* 6, 121–8.

Cane, M.A. 1983 'Oceanographic events during El Niño', *Science* 222, 1189–94.

Caviedes, C. 1973. 'Secas and El Niño: Two simultaneous climatical hazards in South America', *Assoc. Amer. Geogr., Proceedings* 5, 44–9.

Caviedes, C. 1981. 'Rainfall in South America. Seasonal trends and correlations', *Erdkunde* 35, 107–18.

Caviedes, C. 1982a. 'On the genetic linkages of precipitation in South America', in D. Havlik and R. Mackel (eds.) *Fortschritte landschaftsokologischer und klimatologischer Forschungen in den Tropen, Freiburger Geographische Hefte* 18, 55–77.

Caviedes, C. 1982b. 'Natural hazards in South America: In search for a method and a theory', *GeoJournal* 6, 101–10.

Caviedes, C. 1985. 'Fluctuations of the South Pacific anticyclone during the ENSO 1976–77 and 1982–83', *Tropical Ocean-Atmosphere, Newsletter* 29, 3–4.

Chu, P.S., Hastenrath, S. 1981. *Diagnostic studies of Brazil. Preliminary results*, University of Wisconsin, Climate Analysis Center, Madison, USA.

Doberitz, R. 1969. 'Cross-spectrum and filter analysis of monthly rainfall and wind and data in the tropical Atlantic region', *Bonner Meteor. Abhandlungen* 8.

Francou, B., Pizarro, L. 1985. 'El Niño y la sequía en los altos Andes centrales (Perú y Bolivia)', *Bulletin Institut Français d'Etudes Andines* 14, 1–18.

Halpern. D., Hayes, S.P. 1983. 'Oceanographic observations of the 1982 warming of the Tropical Eastern Pacific', *Science* 221, 1173–5.

Lukas, R., Hayes, S.P., Wyrtki, K. 1984. 'Equatorial sea level response during the 1982–1983 El Niño', *J. Geophys. Res.* 89, 10425–30.

Markham, C., McLain, D.R. 1977. 'Sea-surface temperature related to rain in Ceara, Northeastern Brazil', *Nature* 265, 320–3.

Mugica, R. 1984. 'Departamento de Piura rainfall in 1983', *Tropical Ocean-Atmosphere, Newsletter* 28, 7.

Nobre, P., Moùra, A., Nobre, C. 1985. 'Planetary scale circulation anomalies associated with droughts over northeast Brazil', *Tropical Ocean-Atmosphere, Newsletter* 30, 11–13.

Rasmussen, M.E., Hall, J.M. 1983. 'El Niño: The great equatorial warming event of 1982–1983', *Weatherwise* 36, 166–76.

Romero, H. 1986. 'El clima de Chile', *Geografía de Chile*, Santiago de Chile, Instituto Geográfico Militar.

Todorovic, P. 1978. 'Stochastic models of floods', *Wat. Resources Res.* 14, 345–56.

Virji, H. 1981. 'Preliminary study of summertime tropospheric circulation patterns over

South America estimated from clouds and winds', *Mon. Wea. Rev.* 102, 599–610.

Vulquin, A. 1971. 'Argument en faveur d'un mousson en Amazonie', *Tellus* 23, 74–81.

Waylen, P. 1985. 'A method of predicting daily peak flows in the high flow season', *J. Hydrol.* 77, 89–105.

Waylen, P., Caviedes, C. 1986. 'El Niño and annual floods on the northern Peruvian littoral', *J. Hydrol.* 89, 141–56.

Wright, P.B. 1984. 'Relationships between indices of the southern oscillation', *Mon. Wea. Rev.* 112, 1913–19.

Wyrtki, K. 1975. 'El Niño: The dynamic response of the equatorial Pacific to atmospheric forcing', *J. Phys. Oceanogr.* 5, 572–84.

Wyrtki, K. 1979. 'The response of sea surface topography to the 1976 El Niño', *J. Phys. Oceanog.* 9, 1223–31.

Chapter 23

Outline of a historical weather data base for Japan

M. Yoshimura and M.M. Yoshino   Yamanashi University and University of Tsukuba, Japan

Introduction

In order to chart the historical development of the climate in Japan the Historical Weather Reconstruction Group and Information and Computer Centre of Yamanashi University are currently co-operating in the construction of a data base. Descriptions of weather in the period 1700–1880 have been collected on a daily basis from the diaries kept by the local governments (Han) during the Edo era, by the priests of large shrines or temples, by farmers, and so on.

Before we started to input the description of weather data, we discussed the whole planning of the data base at the meetings of our Group. The present paper gives an outline of our data base. We are sure that such a data base will be widely used, and, accordingly, the issue of international standardisation needs to be considered.

Input of description

Considering the users' purpose, the descriptions of each day were extracted from the diaries. Code numbers or signs were determined according to the weather of the day and the variation of weather and its phenomena: for example, fine weather, warm or cold; and cloudy weather, dry or moist. Also, any remarkable phenomena and wind conditions were noted. The input of the original information was done by 'Katakana', which is an expression in Japanese. An example of data description is shown in Figure 23.1, where the note on the right hand side is given by 'Katakana'. For example, for the first and second lines the note says 'cloudy; occasional rain; twice thunder in north at night: then rain'.

The descriptions of weather in the diaries only follow the complete changes of weather during the day in a few cases. On the assumption that fine or clear weather is the best and snow is the worst condition, 11 weather types are set between two weather conditions mentioned above. The weather range of each day is thus shown by a combination of 13 weather types. The best condition is written in the second place and the worst one in the first place.

Figure 23.2 shows the list of signs of the weather. On the left hand side, weather conditions are given on H1 to H10, wind conditions on J and notes on I. At present, input has been finished for 11,000 months of data at a total of 18 points. In future, the number of months will be increased, but not the station numbers to any great extent. Finally, input will be made for about ten points for every day since 1650.

```
        A          B      C        D            E     F    G  H I  J
FKN1807814 FKN1807 17 NJ-53-1214 KANA01 0101 38 B 00 M 0 0 0 クモリ トキドキ アメ ヨナカ キタ
ノ ホウ ライメイ 2ト ソノアト アメ
FKN1807814 FKN1807 17 NJ-53-1214 KANA01 0102 11 N 00 0 0 0 0 セイテン
FKN1807814 FKN1807 17 NJ-53-1214 KANA01 0103 11 N 00 0 0 0 0 ハレ キコウ ハル 2ガツ ノ ゴ
トシ
FKN1807814 FKN1807 17 NJ-53-1214 KANA01 0104 11 N 00 0 0 0 0 カイセイ ヨナカ オボロヅキ シュ
ンヤ ノ コトシ
FKN1807814 FKN1807 17 NJ-53-1214 KANA01 0105 37 G 00 0 4 W 7 アサ 5ツ コロ ビウ ハレテ クモ
リ ニシカゼ サワガシイ 7ツ コロ ヨリ ニシカゼ オオイニ オコル ヨル ニ ハイリ カゼ マスマス ツヨイ 8ツハン コロ ヨリ カゼ ヤム
FKN1807814 FKN1807 17 NJ-53-1214 KANA01 0106 CC N 50 0 0 0 0 ヒセツ ハナハダ サムイ
FKN1807814 FKN1807 17 NJ-53-1214 KANA01 0107 00 0 5I 0 0 0 0 ハナハダ カンキ ツヨイ ススリ
スミ コオル
FKN1807814 FKN1807 17 NJ-53-1214 KANA01 0108 88 N 00 0 0 0 0 アメ
FKN1807814 FKN1807 17 NJ-53-1214 KANA01 0109 88 N 00 0 0 0 0 アメ
FKN1807814 FKN1807 17 NJ-53-1214 KANA01 0110 88 N 00 0 0 0 0 アメ
FKN1807814 FKN1807 17 NJ-53-1214 KANA01 0111 88 N 00 0 0 0 0 アメ
FKN1807814 FKN1807 17 NJ-53-1214 KANA01 0112 88 N 00 0 0 0 0 アメ
FKN1807814 FKN1807 17 NJ-53-1214 KANA01 0113 11 N 00 0 0 0 0 ハレ
FKN1807814 FKN1807 17 NJ-53-1214 KANA01 0114 11 N 00 0 0 0 0 カイセイ キコウ ハル ノ コトシ
FKN1807814 FKN1807 17 NJ-53-1214 KANA01 0115 11 N 20 0 0 0 0 カイセイ アタタカ ニテ ハル ノ コ
トシ
```

A:Data and year code   B:Code of prefecture

C:Mesh code D:Code of place name, city, town or village

E:Date   F:Range of weather G:Tendency of weather

H:Code for additional condition, such as hot or cold, and dry or wet,
,and associated phenomena with such condition

I:Code for remakable phenomena, for example lighning, snow storm, and so on.
J:Code for wind condition

Figure 23.1   An example of data description

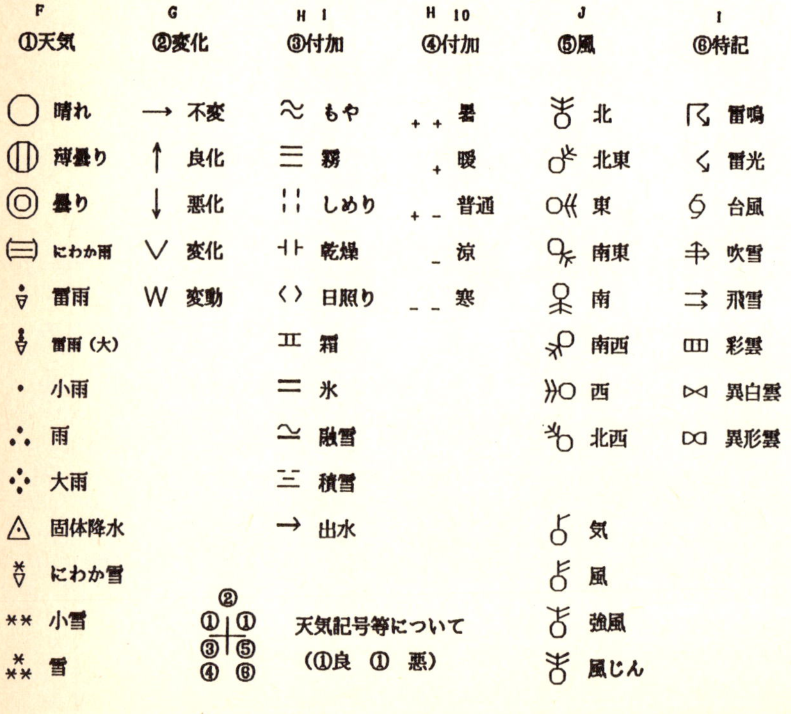

Figure 23.2   List of weather signs

1843 年

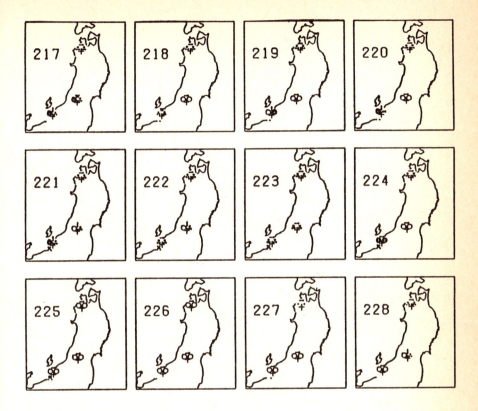

Figure 23.3 An example of split map for Tohoku district, north-east Japan, 17–28 February, 1843

Output of weather information

Figure 23.3 shows an example of a split map for Tohoku District, NE Japan, from 17–28 February 1843. Another example of a map of weather distribution in Japan for the period 1–11 1865, is shown in Figure 23.4. In the split map or distribution map, selection of information for weather should be necessary when the number of stations increases.

In a similar way, 18 kinds of map can be provided. Besides the distribution of the whole area of Japan (Figure 23.4) and the local area (Figure 23.3), sequences of weather at 12 stations over one month can be obtained in one output. The stations are arranged automatically in order from the upper to the lower part on the output as from north-east to south-west Japan, as shown in Figure 23.5.

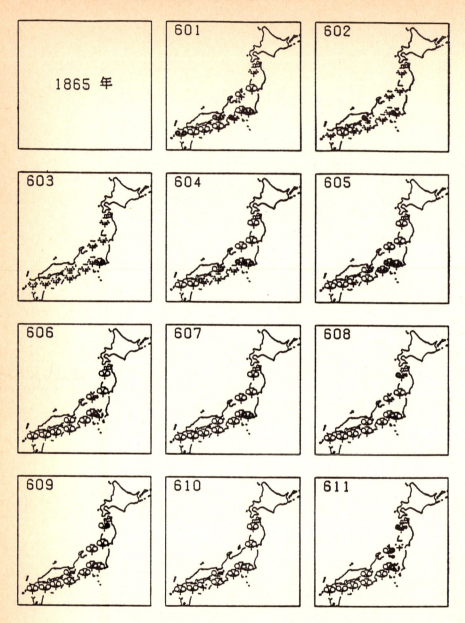

Figure 23.4 An example of weather distribution in Japan, 1–11 June 1865

Concluding remarks

There are three characteristic points in this data base. First, non-protocol terminals, such as personal computers, are chosen, considering convenience to the end users. Second, by using command procedure or macro-command communication, the data base is much simplified. Third, the map data should be also set in correspondence

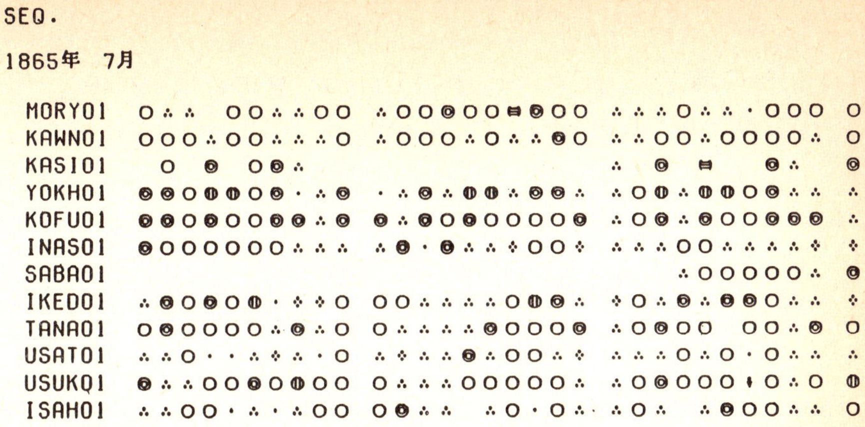

SEQ.

1865年 7月

```
MORYO1   O ∴ ∴   OO∴∴OO   ∴OOⒸOOⒽⒸOO   ∴∴∴O∴∴·OOO  O
KAWNO1   OOO∴OO∴∴∴O   ∴OOO∴O∴∴ⒸO   ∴∴OO∴OOOO∴   O
KASIO1    O  Ⓒ  OⒸ∴                      ∴  Ⓒ  Ⓗ    Ⓒ∴      Ⓒ
YOKHO1   ⒸⒸOⓂOⒸ·∴Ⓒ   ·∴Ⓒ∴ⓂⓂ∴ⒸⒸ∴   ∴OⓂ∴ⓂⓂOⒸ∴∴   ∴
KOFUO1   ⒸⒸOⒸOOⒸⒸ∴Ⓒ   Ⓒ∴ⒸOⒸOOOOⒸ   ∴OⒸ∴ⒸOOⒸⒸ   ∴
INASO1   ⒸOOOOOO∴∴∴   ∴Ⓒ·Ⓒ∴∴∴OO❖   ∴∴∴OO∴∴∴∴❖   ❖
SABAO1                            ∴OOOOO∴   Ⓒ
IKEDO1   ∴ⒸOⒸOⓂ·❖❖O   OO∴∴∴∴OⓂⒸ∴   ❖O∴Ⓒ∴ⒸⒸO∴∴   ❖
TANAO1   OⒸOOOO∴Ⓒ∴O   O∴∴∴∴ⒸOOOⒸ   ∴OⒸOO   OO∴Ⓒ  O
USATO1   ∴∴O··∴❖∴·O   ∴❖∴∴Ⓒ∴OO∴❖   ∴∴∴O∴O·O∴∴   ∴
USUKQ1   Ⓒ∴∴OOⒸOⓂOO   O∴∴∴OOOOO∴   ∴OⒸOOO◆O∴O   Ⓜ
ISAHO1   ∴∴OO·∴·∴OO   OⒸ∴∴  ∴O·O∴·   ∴O∴   ∴ⒸOO∴∴   O
```

Figure 23.5   An example of a weather sequence in July 1865

programme for the terminal to save time in interrelating data for maps.

It will take about two years for reading data and for editorial work on a thesaurus.

Acknowledgement

We would like to express our sincere thanks to Mr Hitoshi Ariizumi, Chief of System Development, Computer Centre, Yamanashi University, for his technical support. This study was supported by grant-in-aid from the Ministry of Education, Science and Culture (Data Base) in 1986.

Chapter 24

Reconstruction of rainfall variation in the Bai-u rainy season in Japan

M.M. Yoshino and A. Murata University of Tsukuba, Japan

Introduction

The Bai-u season, an early summer rainy season in Japan, shows great inter-annual variations both in its duration and the amount of rainfall. Rainfall measurements during this season have been made at several points such as at Matsuzaka and Tsu (Mizukoshi 1983, 1985) and at Hirosaki (Maejima and Tagami 1986) and for the whole of Japan for abnormal years (Maejima and Koike 1976; Yaji and Misawa 1981; Mikami 1987).

In order to reconstruct the distribution of the Bai-u season rainfall in the whole of Japan for every year over a longer period, a historical weather data base (Yoshimura and Yoshino, Chapter 23, this volume) is used. The present chapter is a result of our preliminary study. First, we divided Japan into subregions having common variations of June, July, and combined June and July rainfall by varimax rotated principal component analysis (PCA). Second, rainfall variations were estimated in each sub-region by using the description of weather in the old diaries. Third, results of reconstruction of rainfall variation are presented. Finally, we discuss some problems in the reconstruction technique and compare the results with those in China.

Regional division of June and July rainfall by PCA

The first step in reconstructing Bai-u season rainfall was to make a regional sub-division of Japan based on common variations. To find out these regional variations in detail varimax rotated PCA was employed. The data were rainfall measurements at 50 stations in Japan in the months of June and July for the years 1901–84. The calculation was carried out on the logarithmically transformed values. The rotations were applied on the upper ranks higher than 7, 10 and 9 for June, July and both months combined respectively. Rotation of principal component was employed taking into consideration the problems perceived by Richman (1986).

In Figure 24.1 the distribution of factor loadings obtained by PCA is given. We define the homogeneous regions which have factor loading greater than 0.6. Figure 24.2a shows the result of regional division by factor loadings for June rainfall, with Figure 24.2b doing the same for July rainfall and Figure 24.2c for combined June and July rainfall. Regions so divided show some striking climatic features. For June (Figure 24.2a), the contrast between the Pacific coastal areas (regions $F_6 7$, $F_6 2$ and $F_6 3$) and Japan Sea coastal areas (regions $F_6 1$ and $F_6 5$), both including the Pacific and Japan Sea coasts, which may be caused by an indistinct Bai-u rainfall in June. In July, the regions spread systematically from south-west to north-east as

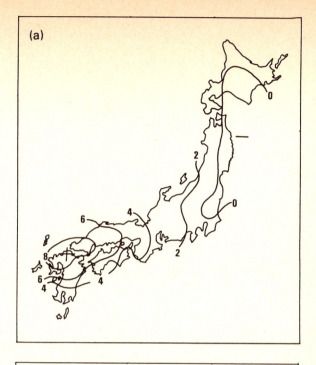

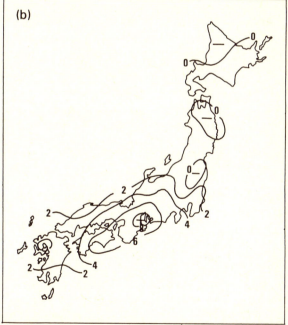

Figure 24.1   Varimax rotated factor loading

(a) First component
(b) Second component

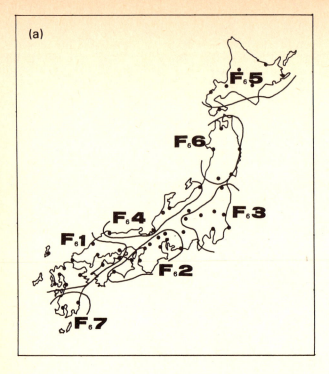

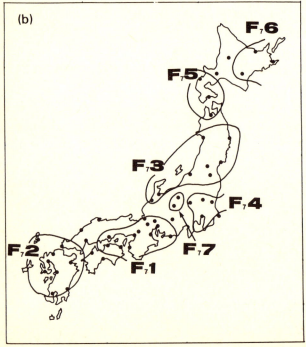

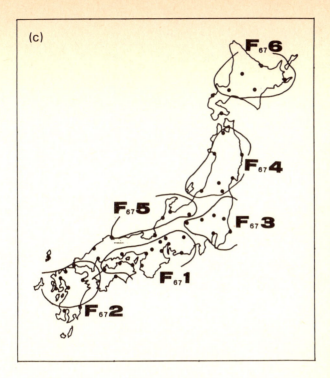

Figure 24.2   Regional division by factor loading for June and July rainfall and their combination

(a) June rainfall
(b) July rainfall
(c) Combined June and July rainfall

is shown in Figure 24.2b. For combined June and July rainfall, as shown in Figure 24.2c, the distribution of regions shows some overlapping of June and July, but seems more representative of the Bai-u conditions: Region $F_{67}1$ is a typical region receiving Bai-u rainfall in this season. The development of the Bai-u front is affected strongly by the activity of the Ogasawara high, a part of the North Pacific anticyclone. Region $F_{67}2$ is the most sensitive region for the moist south-westerly winds during the Bai-u season and occasionally suffers very severe rainfall. $F_{67}3$ is the central part of Japan, where the effect on the rainfall of cold, wet north-easterly winds from the Okhotuk high is observable in addition to Bai-u front activity. Region $F_{67}4$ is the Tohoku district, where the Bai-u season rainfall is affected strongly by the so-called 'Eurasian polar frontal zone' (Yoshino 1965; 1966). Region $F_{67}5$ is the Japan Sea coastal area of Honshu, where heavy rainfall does occur occasionally in the last stage of the Bai-u season. Region $F_{67}6$ is Hokkaido, where the Bai-u rainfall pattern is unclear.

Table 24.1 Percentage of variance explained by the first and second component for each region

Region	P1	P2
$F_6 1$	74.4	9.5
$F_6 2$	73.3	8.8
$F_6 3$	66.1	10.7
$F_6 4$	76.8	9.9
$F_6 5$	61.6	14.5
$F_6 6$	66.8	17.8
$F_6 7$	89.2	10.8
$F_7 1$	70.9	6.7
$F_7 2$	74.6	9.7
$F_7 3$	69.4	11.8
$F_7 4$	69.6	20.3
$F_7 5$	64.9	25.7
$F_7 6$	69.6	17.8
$F_7 7$	88.8	11.2
$F_{67} 1$	71.6	8.5
$F_{67} 2$	73.5	11.0
$F_{67} 3$	68.5	11.4
$F_{67} 4$	68.5	11.4
$F_{67} 5$	76.4	11.4
$F_{67} 6$	63.3	15.1

Table 24.2 Regression equations for each region

Region	Regression equation	R^2
$F_6 3$	$RVI = 0.30 \cdot RD - 3.11$	0.51
$F_7 1$	$RVI = 0.24 \cdot RD - 2.44$	0.57
$F_{67} 1$	$RVI = 1.21 \cdot RD - 12.57$	0.64

Key: RD = Number of days with more than 1 mm rainfall.
　　　 RVI = Rainfall variation index

Estimation of rainfall variation by region

The next step in our study was to employ non-rotated PCA to find the variables which represent well the variations in each of the respective regions above. Because the first component explains the largest part of variations in each region, as given in Table 24.1, we used the score of the first component as the variables for our purpose. This variable is named the Rainfall Variation Index (RVI).

To estimate RVI in the historical time, we used the proxy data, the description of weather in the old diaries. These are now in the 'Historical Weather Data Set' (Yoshimura and Yoshino, Chapter 23, this volume). From the data, monthly total numbers of days with rainfall were summed up. Estimation was made possible by the linear regression equations as given in Table 24.2. The equations given here are preliminary ones, even though the correlation is significant at high levels. Based on a longer data set, more correct equations will be presented in a future paper.

Reconstruction of historical rainfall variations

Figure 24.3a shows the result of the reconstructed series of RVI for the years 1710–1895 in region $F_6 3$, Tokyo and its surrounding area in central Japan. Figure 24.3b shows a comparison of estimated RVI and calculated RVI for the period 1909–42. Figure 24.3c confirms the correlation between the observed and estimated RVI values in Region $F_6 3$. Roughly speaking, estimated values have a deviation about 1.50, which is equivalent to about four days with rainfall. Using these reconstructed RVI values, long-range variations and their periodicities will be analysed.

Discussion and remarks

Problems in the reconstruction

For estimating RVI, the number of days with more than 1 mm rainfall was used in the present study. It will be tested in detail to see how well this represents the description of 'rain' in the diaries giving rainfall under the period of modern meteorological observation. Does it better represent rainfall of more than 5 mm/day or even more than 10 mm/day?

As has been discussed by Yoshimura (1986), the accuracy of description in the diary is essential. Rainfall during the Bai-u season is particularly local. The different 'rain' descriptions contain both the local differences of rainfall phenomena and the personal differences in describing 'rain'. This must be considered further.

Relationship between Japan and other parts of east Asia

Since the Bai-u in Japan and Mai-yu in China are closely related, their inter-annual rainfall variations have been found in similar phase in the regions under the Pacific Polar Frontal Zone (PPFZ), as has been described by Yoshino (1963; 1977; 1978). Recently, after testing by cluster analysis, a regional division was tried using the rainfall data at more than 150 observation points distributed in east Asia for the period 1956–80 (Yoshino and Aoki 1986). The results revealed that there are three main regions which had similar inter-annual variations in June and July. Among them, a zone extending along the Bai-u frontal zone is located only over Japan in June, but it enlarges strikingly from the lower Yangtze River region, including the southernmost part of the Korean peninsula, to the whole of Japan except Hokkaido in July.

In the present chapter, we tried to check such regionality further in detail, calculating correlation coefficients between the rainfall amounts of each of the regions shown in Figures 24.2 and those at each station in China and the Korean peninsula. Examples from the interesting results are shown in Figure 24.4. Region $F_7 1$, which includes the eastern areas of Shikoku and Chugoku, the main part of Kinki and the western part of central Japan, as shown in Figure 24.2b, has significantly positive correlation with the region extending from the lowest course of the Yangtze River to the southernmost part of the Korean peninsula as shown in Figure 24.4a. Region $F_7 2$, Kyushu and the westernmost part of Chugoku, as

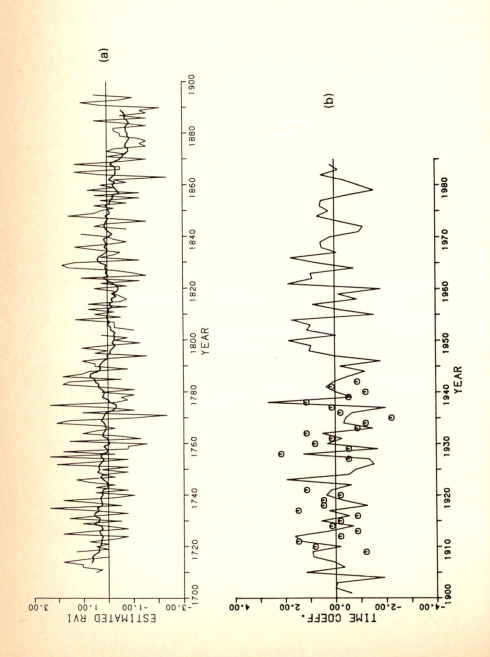

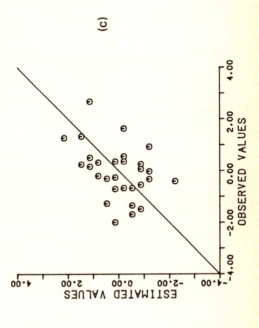

(c)

(a) Secular variation of the RVI estimated by the number of days in the diaries at Yokyo (Yokohama) for 1710–1895 in region F_63

(b) Secular variation of the RVI calculated by the instrumentally observed rainfall in Region F_63 (thick line). Circles show the RVI estimated by the observed number of days with more than 1 mm rainfall at Tokyo.

(c) Observed RVI and estimated RVI in Region F_6.

Figure 24.3 Secular variation, observed and estimated RVI in Region $F_6$3

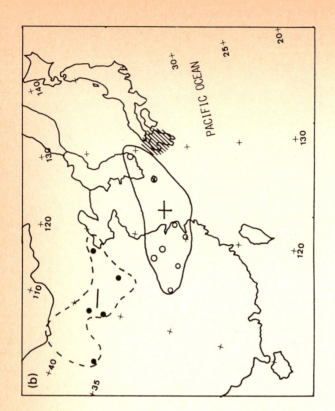

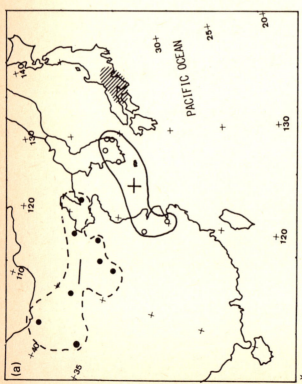

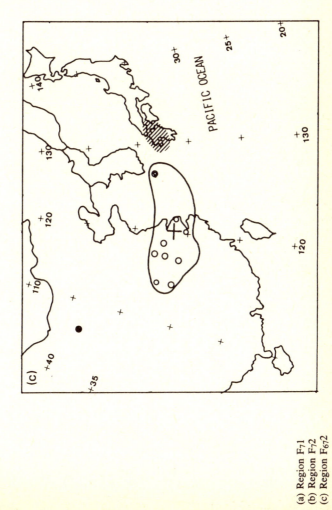

Figure 24.4 Distribution of significantly correlated stations in east Asia*

(a) Region F_71
(b) Region F_72
(c) Region F_672

* White circle denotes positive correlation, black circle negative. Large circle shows correlation significant at the 1% level, small circle at 5%.

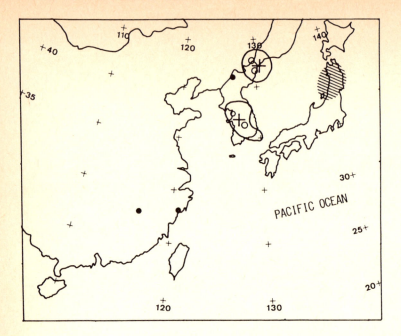

Figure 24.5   Distribution of significantly correlated stations in east Asia with region
$F_{67}4*$

* White circle denotes positive correlation, black circle negative. Large circle shows correlation significant at the 1% level, small circle at 5%.

shown in Figure 24.2b, also has similar characteristics (see Figure 24.4b). This regionality corresponds almost perfectly to the region in July detected by cluster analysis mentioned above. It is worth noting that there appear significant negative correlation regions extending from Mongolia to east China, as shown in Figures 24.4a and 24.4b. These were not shown in the previous studies. They are considered to be related to the activity of the eastern part of the Eurasian Polar Frontal Zone (EPFZ). Another example for region $F_{67}2$, given in Figure 24.4c, reveals that (i) the positively correlated region appears in the lower Yangtze River as in the case of July, (ii) the negatively correlated region becomes very small in Mongolia, and (iii) the positively correlated region is located in the eastern part of the Tibetan plateau. These regionalities will be studied synoptically in the future.

In contrast, region $F_{67}4$, Tohoku, the north-eastern part of Honshu, shown in Figure 24.2c, has close positive relations with some stations in the Korean peninsula, based on the inter-annual variation of combined June and July rainfall, as given in Figure 24.5. This zonally correlated region has been revealed by previous studies (Yoshino 1963; 1965; 1966) by analysing the EPFZ under the influence of the flow pattern at the 500 mb level.

Relationship between rainfall and air temperature

In regions $F_{6}3$, $F_{7}1$, $F_{7}2$, $F_{7}4$, $F_{67}1$, $F_{67}2$ and $F_{67}3$ a negative correlation is mostly found between air temperature and RVI. This suggests that it might be possible to reconstruct air temperature. Roughly speaking, if air temperature is low, RVI is high, indicating the Little Ice Age condition. As Wang *et al.* (1987) stated, floods in the lower Yangtze River region were more frequent during the Little Ice Age. These should be studied further.

Conclusion

A reconstruction of rainfall during the Bai-u season was attempted. First Japan was divided into homogeneous regions having a common inter-annual variation on the basis of recent instrumental observation. Second, the rainfall variation index was calculated for each region so defined. Third, experimental equations were obtained to estimate the rainfall variation index from the description of rain in the old diary sources for each region. Fourth, based on the results reconstructed in this way, the correlated zonal areas in east Asia under the influence of Polar Frontal Zones were discussed and further problems to be studied were outlined. This is a preliminary report, but it is suggested that such an approach will provide considerable information about the historical pattern the Bai-u rainfall.

Acknowledgement

This chapter is a part of the results of a study on the 'Reconstruction of past climate in the WCRP in Japan', 1987.

References

Maejima, I., Koike, Y. 1976. 'An attempt at reconstructing the historical weather situation in Japan', *Geogr. Repts. Tokyo Metropolitan Univ.* 11, 1–12.

Maejima, I., Tagami, Y. 1986. 'Climatic change during historical times in Japan', *Geogr. Repts. Tokyo Metropolitan Univ.* 21, 157–71.

Mikami, T. 1987. 'Climate of Japan during 1781–90 in comparison with that of China' in Ye Duzheng *et al.* (eds), *The Climate of China and Global Climate*, Beijing, China Ocean Press, and Berlin, Springer-Verlag, 63–75.

Mizukoshi, M. 1983. 'Weather and climate in the southern part of the Ise plain during the second half of the eighteenth century, *Kisho-kenkyu Note* 147, 99–106 (in Japanese).

Mizukoshi, M. 1985. 'Reconstruction of the climate during the period 1831–40 in Kinki-Tokai district, central Japan', *Ann. Rep. Inst. Disast. Prev.*, Kyoto Univ., 28, B-2, 121–32 (in Japanese).

Richman, M.B. 1986. 'Rotation of principal components', *J. Climatol.* 6, 293–335.

Wang, Shaowu *et al.* 1987. 'Drought/flood variations for the last 2000 years in China and comparison with global climatic change' in Ye Duzheng *et al.* (eds) *The climate of China and Global Climate*, Beijing, China Ocean Press, and Berlin, Springer-Verlag, 20–9.

Yaji, M., Misawa, A. 1981. 'A study on the climate around the Tempo famine', *Sci. Repts. Yokohama Natl. Univ., Sec. II* 28, 91–107 (in Japanese).

Yoshimura, M. 1971. 'Regionality of secular variation in precipitation over Monsoon Asia and its relation to general circulation' in M.M. Yoshino (ed.) *Water Balance of Monsoon Asia*, Tokyo, University of Tokyo Press, 195–215.

Yoshimura, M. 1986. 'Reconstruction of climate in the Edo period' in T. Kawamura (ed.) *Periodicity and regionality of climatic change*, Tokyo, Kokinshoin, 226–43 (in Japanese).

Yoshino, M.M. 1963. 'Rainfall, frontal zones and jet stream in early summer over east Asia', *Bonner Met. Abh.* 3, 1–126.

Yoshino, M.M. 1965. 'Four stages of the rainy season in early summer over East Asia, I', *Jour. Met. Soc. Japan* 43, 231–45.

Yoshino, M.M. 1966. 'Four stages of the rainy season in early summer over east Asia, II', *Jour. Met. Soc. Japan* 44, 209–17.

Yoshino, M.M. 1977. 'Bai-u, the rainy season in early summer' in E Fukui (ed.) *Climate of Japan*, Amsterdam, Elsevier, 85–101.

Yoshino, M.M. 1978. 'Regionality of climatic change in Monsoon Asia' in K. Takahashi and M.M. Yoshino (eds), *Climatic Change and Food Production*, Tokyo, University of Tokyo Press, 331–42.

Yoshino, M.M., Aoki, T. 1986. 'Interannual variations of summer precipitation in east Asia: Their regionality, recent trend and relation to sea-surface temperature over the North Pacific', *Erdkunde* 40, 94–104.

Chapter 25

Recent climatic change in Australasia

J.E. Hobbs University of New England, Australia

Introduction

The region considered in this chapter includes Australia, New Zealand and Papua
New Guinea. Most previously reported work relates to Australia and New Zealand,
with very little work having been done in Papua New Guinea, mainly because of
the inadequacy of climatic records. The emphasis in the chapter is on rainfall
changes rather than temperature changes, again because rainfall records are
generally longer, more complete, and more readily available, for a greater range
of stations.

The Australasian region includes areas dominated by summer rainfall, areas
dominated by winter rainfall, and areas with rainfall more evenly distributed
throughout the year, with the possibility of winter maxima in some years and
summer maxima in others. Broad patterns of mean atmospheric circulation in the
region demonstrate the overall effects on rainfall of the subtropical high pressure
zone, of the intertropical convergence zone in lower latitudes, and of the mid-
latitude westerlies. The latitudinal and longitudinal extents of the region are such
that it is unrealistic to expect similar patterns of change throughout. Therefore, the
detailed picture of recent climatic change, particularly of rainfall change, and the
possible explanations for such change, are complex.

The work reported in this paper includes preliminary results of an analysis of
monthly rainfall data for the two Australian mediterranean-type climatic regions in
Western Australia and South Australia. These have been selected partly because of
the expectation that their annual rainfall totals should show strong seasonal
dominance of winter rainfall (this is what might be regarded as the conventional
wisdom), partly because they are longitudinally separated and so might be expected
to show characteristics of change towards opposite sides of the continent, and partly
because of the possibility of subsequent comparisons with other similar climatic
regions in both hemispheres.

Evidence from the instrumental record

Australia

The few station records back to the 1800s suggest that much of the eastern half of
Australia was wetter in the nineteenth century than in the first half of the twentieth
century. There seems to have been a return to higher annual rainfalls over much
of eastern Australia around 1945–6 (Pittock 1975). Annual rainfalls have since
increased by 10–20 per cent over a large area of south-east Australia, although it

is interesting to note Pittock's estimate that the area of significant (at the 95% level) rainfall change in Australia (1941–74 means compared with 1913–40 means) is roughly what would be expected from a random data set.

Short-term and medium-term changes in rainfall, with a break in the early to mid-1940s, have been identified by several researchers, for example, Hobbs (1971, 1972), Pittock (1975), Cornish (1977) and Russell (1981). Pittock (1983) updated his earlier analysis and examined the changes in terms of three-month seasons. His analysis pointed to a winter decrease in mean rainfall in south-western Australia and on the western slopes of the Divide in New South Wales; an increase in excess of 20 per cent in spring in most of inland New South Wales and parts of South Australia and Victoria, with a maximum in excess of 30 per cent in central New South Wales; a summer increase in excess of 20 per cent in parts of north-central New South Wales; and large areas in western new South Wales, central-southern Queensland and parts of South Australia showing autumn increases up to 40 per cent with much of inland Western Australia having a decrease. These data support the increase in annual rainfall reported in Pittock (1975) and point to its concentration in spring, summer and autumn, which in turn points to a change in pattern of seasonal distribution of rainfall. Russell (1981), on the basis of an analysis for the period 1895–1974, also identified a trend of increasing summer rainfall over much of south-eastern Australia. His analysis found some patterns of discontinuity in summer and winter rainfalls over short and medium-length periods. Pronounced discontinuities of summer rainfall occurred in 1944–6 and in winter rainfall in 1950 over large areas of Australia. The period 1946–55 was identified as one of above average summer rainfall (October to March) and 1950–65 was identified as a period of above average winter rainfall in some areas.

Cornish's (1977) analysis of changes in seasonal and annual rainfall was confined to New South Wales. His work indicates a significant increase in annual and summer rainfall in central New South Wales since 1945, with annual increases in some parts of up to 20 per cent, and summer (January to March) increases of up to as much as 50 per cent but generally of between 20 and 30 per cent. His results were consistent with those identified by Hobbs (1972) of substantial changes in mean summer rainfalls since 1935 in the area of north-eastern New South Wales.

Kraus (1954) found a decrease in summer rains in eastern Australia to a minimum near the turn of the century with a subsequent increase in the early part of the twentieth century. He found a negative correlation between these summer rainfall trends and the strength of the upper westerly winds, and indicated that the twentieth-century increase in rainfall occurred in phase with a decrease in the 300 hPa wind speed. Cornish (1977) suggested that the zone of summer rainfall dominance had extended further south due to changes in southern hemisphere pressure patterns. Similarly, Ward and Russell (1980) suggested that greater developments of the Queensland heat trough in summer contributed to increased summer rainfall after 1935 in New South Wales, south-east Queensland and eastern South Australia.

Deacon (1953) found that the most significant increases in summer rainfall from the period 1891–1910 to the period 1911–40 were in the Adelaide region of South Australia, where summer rainfall is much lower and more irregular than in the eastern states. He pointed to a possible increase in activity of meridional troughs between anticyclones, thus increased meridional exchanges. Russell (1981) and Iwasaki (1985) identified two different zones of rainfall trends in the Eyre peninsula of South Australia. The north-eastern zone of the peninsula displayed trends of summer rainfall similar to parts of eastern Australia, with a slight decrease to about 1950, and then a steady increase; while the coastal zone of the peninsula showed an increasing rainfall trend to the mid-1950s followed by a slight decrease. This suggests that the Eyre peninsula, which is part of the South Australia mediterranean region, might be influenced by trends in both western and eastern Australia, at least in summer. It is possible that the South Australian region might exhibit characteristics transitional between the western and eastern sides of the Australian continent.

Gentilli (1971) and Wright (1974) both described an increase in late summer rainfall from the period 1881–1910 to the period 1911–40 in Western Australia, possibly due to an increased frequency of tropical cyclone activity in the Timor Sea. They also found an increase in Western Australian winter rainfall early in the twentieth century, with the month of highest rainfall shifting from July in 1877 to June around 1950. The suggested general circulation changes included a slight westward shift of the quasi-permanent trough near 110 °E in winter and an eastward shift of long wave patterns this century in late winter, decreasing the influence of showery westerly airstreams in Western Australia.

Ward and Russell (1980) established that east coast winter rainfalls were heavier before 1935 than afterwards, a reversal of apparent trends in South Australia and Victoria. They suggested that anticyclonic conditions were longer-lasting over the eastern Great Australian Bight and New South Wales, and less developed over the western Bight before 1935 than after.

At least part of the autumn and winter rainfall in South Australia comes with north-west cloud bands, which form ahead of long-wave troughs in the Western Australian region. Wright (1974) inferred that a westward movement of the average position of west coast troughs may have caused a decrease of such rainfall in South Australia during early winter, corresponding with possible increases observed in Western Australia at the same time.

A recent study by Hobbs (1987) of critical season rainfalls (i.e. rainfall periods considered by agronomists to be of critical significance for wheat production) for the Australian wheatbelt, which extends from central Queensland through inland New South Wales, northern Victoria, South Australia and into the south-west of Western Australia, has identified marked shifts in the temporal and spatial patterns of rainfall in the period 1891–1980. For each of the major regions of the wheatbelt the one-year-in-ten rainfall amounts were mapped for four periods: the full record and three 30-year periods, 1891–1920, 1921–50 and 1951–80. This permitted identification of areal shifts of risk through time.

In the summer rainfall region of central Queensland indications were of a westward shift of isohyets from the earliest to the most recent period, suggesting decreasing risk, as the amounts of rainfall expected once in ten years have increased. In the region of southern Queensland and northern New South Wales, also experiencing summer rainfall, the signs were of a two-way shift about a north–south axis running through the centre of the region, with western areas tending towards slightly decreased risk and eastern, especially north-eastern, areas tending towards increased risk. In the region of central and southern New South Wales, Victoria and South Australia, where rainfall grades from an evenly distributed amount through the year to a winter maximum, variations between the successive 30-year periods were not marked, although there was some indication of an inland shift of higher rainfall amounts, mainly through the western part of the region in New South Wales and in northern Victoria. In the winter rainfall region of Western

Australia there was evidence of a westward shift of all isohyets, pointing to increasing risk.

Information about possible temperature changes during the period of instrumental records is less readily available, because of the relative paucity of temperature data for long periods. On a hemispheric basis, the increase in annual mean temperature between the 1910s and the 1970s has been about 1.3 °C in the 30–50 °S zone (Paltridge and Woodruff 1981). The only significant results from a study of yearly average spring temperatures (Tucker 1980) were positive trends in daily minimum temperatures for four stations in South Australia. Coughlan (1979) found that between 1946 and 1975 most of Australia experienced a rise in annual mean maximum temperatures, with the greatest increases in inland south-eastern Australia.

New Zealand

Trenberth (1977) reported that changes in New Zealand rainfall have not been as pronounced as those in temperatures, although there has been a general increase in the northern area and a decrease in the Canterbury area. Seelye (1950) detected a declining trend in annual rainfall in New Zealand from 1855 to 1940, with some recovery in the following decade. These trends could have been related to a decrease in spring westerly airstream rainfall from 1912 to 1950 identified for the west coast of the South Island by De Lisle (1956). He inferred a general poleward displacement of the spring position of subtropical anticyclone tracks during that period.

Minor glacial advances have been identified for New Zealand in the period 1890–1920, in the 1930s, in the early 1950s, and in the mid-1960s. The general warming until 1940 in the northern hemisphere was also present in the southern hemisphere, but there has been little evidence of subsequent cooling. In New Zealand temperatures actually rose quite sharply after about 1950, with a small decrease in the mid-1950s corresponding with the minor glacial advance. Such rises were evident at both rural and urban stations (Trenberth 1977). There has been a trend for more north-easterly or less south-westerly flow across New Zealand largely due to an increased tendency for anticyclones to persist east of New Zealand rather than over Australia and the Tasman Sea, and to their preference for higher latitudes. The changes identified in New Zealand temperatures and rainfall could be accounted for at least in part by these changes in atmospheric circulation.

Papua New Guinea

Apart from Australia and New Zealand there is little evidence available for climatic change elsewhere in Australasia. Studies are hampered by lack of data and by the fact that very few stations have records for longer than 30 years, with many of these being of doubtful reliability. Analysis of rainfall trends at Port Moresby for the period 1945–6 has shown a gradual increase in rainfall during the north-west season (summer), which in turn contributes significantly to a rise in annual rainfall. This increase in annual rainfall has been particularly apparent since the late 1950s (Magari 1980).

Australian mediterranean regions

The data set for this investigation comprised monthly rainfall totals for seven stations in the mediterranean region of Western Australia and 19 stations in the corresponding region of South Australia (see Figure 25.1). In all cases the records were complete for at least 84 years up to and including 1983. Analysis of the data was confined to consideration of winter (May–October) rainfalls only.

The temporal patterns of winter rainfall variation were summarised using an index of normalised May–October rainfall departures (Lamb 1985). The value of the index, R, for the year j, is given by

$$R_j = \frac{1}{N} \sum_{i=1}^{N_j} \frac{r_{ij} - \bar{r}_i}{\sigma_i}$$

where r_{ij} is the May–October rainfall for station i in the year j; $\bar{r}_i$ and σ_i are respectively the mean and standard deviation of the May–October rainfall for that station; and N_j is the number of stations with complete records in year j. For the seven Western Australian stations the period of record used was 1900–83 and for the 19 South Australian stations 1891–1983 (see Figure 25.2).

Cumulative departures from the long-term mean winter rainfalls were also calculated for each station (see Figure 25.3).

The proportions of annual rainfall received in the May–October period have varied considerably between the two mediterranean regions. The decadal averages for three representative stations in Western Australia and nine representative stations in South Australia are given in Table 25.1. The Western Australian region is clearly one with a more marked winter dominant rainfall pattern with all decades for the three stations having above 70 per cent of their annual rainfall in winter, most above 80 per cent. On the other hand, for the South Australian region, none of the stations had a single decade with above 80 per cent winter rainfall, and many decades were in the 50–70 per cent range. This emphasises the fact that the South Australian region, although still dominated by winter rainfall, is closer to the summer rainfall influences from the north-east of the continent and is in an area of greater inter-annual variability. It also raises some possible doubt about the legitimacy of our traditional labelling and consideration of the mediterranean regions of Western Australia as similar climatic regions.

The data in Table 25.1 also point to some variability between decades at individual stations. In the case of the Western Australian stations winter rainfall dominance was generally less marked in the last decade of the period than in the first, although the variation through the decades did not conform to any overall pattern. For station 9018 a general decrease is apparent in the proportion of winter rainfall; for station 9619 the last three decades of the period show lower proportions of winter rainfall, with a secondary trough in the decade 1911–20; and for station 10647 there were clear troughs in the decade 1911–20 and 1951–60.

The picture for the South Australian stations also seems to be one of irregular patterns of decadal variations of winter rainfall. Most stations show a smaller percentage of winter rainfall in the decade 1971–1980 than in the decade 1891–1900, but the differences are not great. For most stations in this region between-decade variations are apparent, but there is little correspondence between the temporal patterns of variation in the two regions.

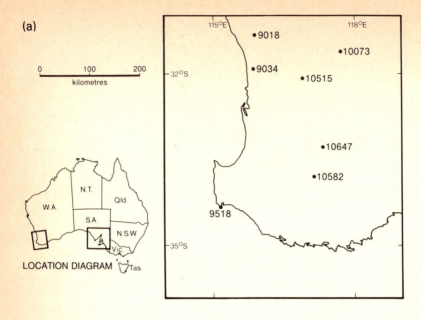

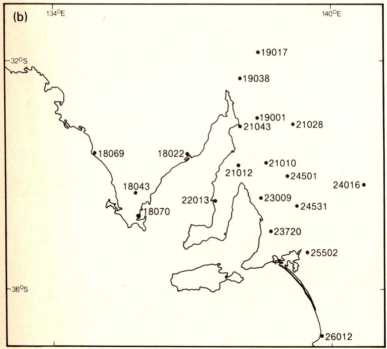

(a) Western Australia
(b) Southern Australia

Figure 25.1 The mediterranean regions of Western Australia and South Australia, showing locations of rainfall stations used in the analyses

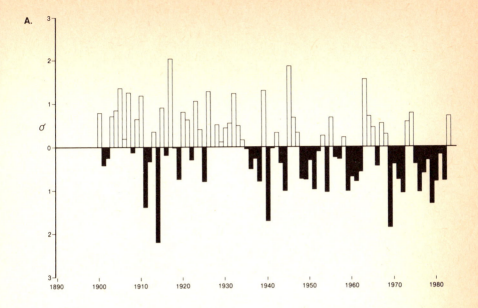

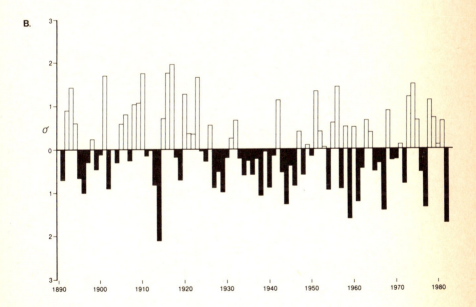

(a) Western Australia
(b) South Australia

Figure 25.2   Time series of yearly averages of normalised winter rainfall departures for Western Australia and South Australia

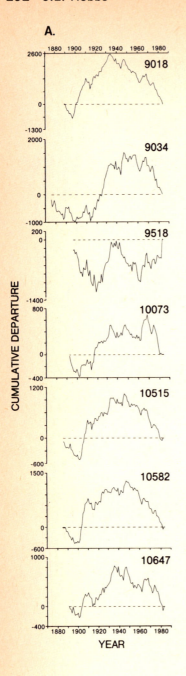

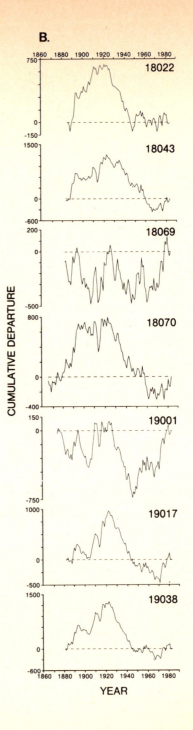

B(continued)

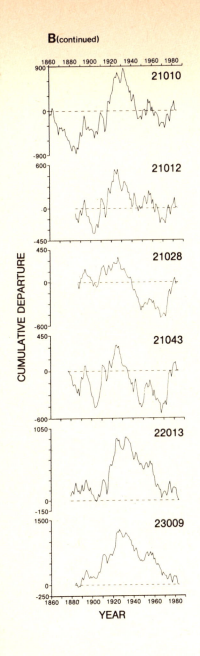

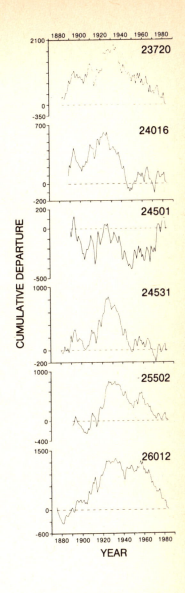

(A) Western Australia
(B) South Australia

Figure 25.3 Cumulative departures from long-term means of winter rainfall for stations in Western Australia and South Australia

Table 25.1 Decadal average percentages of annual rainfall received in winter (May–October) at stations in Western Australia and South Australia

Station No.	1890–1900	1901–10	1911–20	1921–30	1931–40	1941–50	1951–60	1961–70	1971–80
Western Australia									
9018		93	87	87	85	86	84	83	85
9619		85	81	82	85	81	79	78	79
10647		81	73	80	76	80	74	79	73
South Australia									
18022	68	60	63	61	57	53	60	57	63
18069	79	77	77	79	71	75	76	77	77
19017	66	69	59	66	58	55	61	64	58
21012	70	71	70	72	64	62	68	64	69
22013	70	69	71	71	66	64	69	67	70
23009	66	69	69	70	63	62	68	61	65
23720	72	73	75	76	69	71	72	70	70
24016	56	67	56	63	56	54	63	54	58
26012	68	73	74	75	68	68	70	72	71

Comparison of the data in Table 25.1 with the index variations shown in Figure 25.2 does point to a possible association between relatively dry winters and periods when the proportions of annual rainfall occurring in winter were also low.

Temporal patterns of winter rainfall variation

Extremely dry winters occurred in Western Australia in 1911, 1914, 1940, 1969 and 1979. Relatively wet winters were in 1917, 1945 and 1963. Major dry periods occurred between 1907 and 1914, from 1935 to 1944, from 1948 to 1962 and from 1969 to 1982, with only minor breaks in each dry period. Relatively wet periods were in the first decade of this century, from 1915 to 1934 and from 1963 to 1968. Consideration of the whole period 1900–83 indicates predominantly positive departures of winter rainfall before 1935, predominantly negative departures since then. It is interesting that this broad pattern hinges around the mid-1930s, which is close to the time also identified in other analyses of climatic trends as one of a pivotal nature elsewhere in Australia. In a region where winter rainfall dominates, this pattern of winter variation is also reflected in annual totals. In this regard the Western Australian region trends tend to run counter to those which have been observed across the other side of the continent. This is not entirely unexpected, as even casual examination of records of droughts and floods does suggest that major droughts in the east usually accompany relatively wet, often flood, periods in the west and vice versa.

Greater variability is apparent in the time series of the normalised rainfall departures for South Australia. There has been a higher frequency of relatively dry or relatively wet winters, so that extreme dry winters occurred in 1914, 1944, 1959, 1961, 1967, 1977 and 1982. Of these only 1914 was also dry in Western Australia, although interestingly in both regions this was clearly the driest winter in the periods analysed. Relatively wet winters in South Australia were, like Western Australia, more common in the first part of the record, occurring in 1893, 1901, 1910, 1916, 1917, 1921, 1924, 1951, 1956 and 1974. As with the dry winters, there is little correspondence between the wet winters in the two regions apart from 1917.

Major dry periods in South Australia occurred from 1895 to 1900, from 1911 to 1914 and from 1924 to 1950. Apart from the first of these periods these all overlap the major dry periods identified for Western Australia. Relatively wet periods in South Australia again matched Western Australia in the first decade of this century, and partially overlapped one of the others, occurring from 1915 to 1923. This wet period was considerably shorter than the corresponding period in Western Australia. From the early 1950s onwards there has been an alternation of short periods, two- to three-years in length of relatively dry and relatively wet winters, although accumulated values of the index suggest that the period from about 1956 to 1972 was relatively dry, with the periods from 1951 to 1958 and from 1973 to 1981 being basically wet.

The period from 1891 to 1982 in South Australia essentially shows a different and more complex pattern of variation than that in Western Australia. The major trends seem to hinge around the mid-1920s rather than the mid-1930s, together with a second pivotal point in the late 1940s. The overall period to the mid-1920s can be characterised as relatively wet, the period from the mid-1920s to the late 1940s as relatively dry, and the period since then as one of greater variability with shorter wet and dry spells. This does tend to reinforce the view that South Australia sits in a transitional situation, reflecting some elements of trends to the west, others of trends to the east.

Examination of the cumulative rainfall deviations from the long-term means (Figure 25.3), reinforces the basic points identified in the preceding discussion. Clearly, the variation between stations is more marked in the South Australian region. With the exception of station 9518, Cape Leeuwin, where maritime influences dominate, all Western Australian stations emphasise the differences already identified between that region and Western Australia and indicate the greater within-region variation apparent for South Australia.

Conclusions

There is evidence of widespread variation in annual and seasonal rainfall over most of Australasia during the period of instrumental record. The picture of variability for Australia is complex in both time and space, but this is not unexpected in view of the size of the continent. The mediterranean climatic regions examined in this paper both show considerable rainfall variability on apparently irregular time scales. The major variations in the two regions have been out of phase with each other, which may in fact make them easier to explain when explanations are sought.

The evidence for any sustained long-term climatic changes, at least as far as rainfall is concerned, is unclear. The variability present in the period of instrumental record, with relatively dry and relatively wet spells occurring over periods of up to 30 or 40 years, but usually for much shorter periods, may be representative of the 'normal' variability in the much longer term. It is not known whether the period of instrumental record is 'normal' in relation to the longer term, or whether it has been an unusually wet or unusually dry period in this respect.

The search for cycles in Australian rainfall goes back to the first half of the nineteenth century. By the turn of the century there were claims for drought cycles of two, seven, nine, ten, eleven, twelve and nineteen years but work since then has done little to substantiate any of these claims. The evidence presented in this chapter could be seen as supporting Dury's (1983) view that sequences of precipitation in pastoral Australia, and perhaps over much wider areas, can be described by square waves, so that their variation is step-functional. Then blocks of years with high average values alternate with blocks of low average values, but there is rarely any regular periodicity apparent in this alternation. Such seems to be the case for the winter rainfalls of the two Australian mediterranean regions.

Some recent studies have shown that there may be promise for forecasting Australian droughts, at least a few months ahead, in terms of relationships with the ENSO phenomenon. Comparison of the index of normalised rainfall departures used in this study with indices of the Southern Oscillation and the occurrence of El Niño events may throw further light on such possible relationships for some parts of the continent.

References

Cornish, P.M. 1977. 'Changes in seasonal and annual rainfall in New South Wales, *Search* 8, 38–40.

Coughlan, M.J. 1979. 'Recent variations in annual-mean maximum temperatures over Australia', *J. R. Met. Soc.* 105, 707–19.

Deacon, E.L. 1953. 'Climatic change in Australia since 1880', *Aust. J. Physics* 6, 209–18.

De Lisle, J.F. 1956. *Secular Variations of West Coast Rainfall in New Zealand and Their Relation to Circulation Changes*, Wellington, NZ Meteorological Service.

Dury, G.H. 1983. 'Step-functional incidence and impact of drought in pastoral Australia', *Aust. Geogrl. Studies* 21, 69–91.

Gentilli, J. (ed.) 1971. *Climates of Australia and New Zealand, Vol. 13, World Survey of Climatology*, Amsterdam, Elsevier.

Hobbs, J.E. 1971. 'Rainfall regimes of northeastern New South Wales', *Aust. Met. Mag.* 19, 91–116.

Hobbs. J.E. 1972. 'An appraisal of rainfall trends in northeast New South Wales', *Aust. Geogrl. Studies* 10, 42–60.

Hobbs, J.E. 1987. 'Climatic patterns and variability in the Australian wheatbelt' in M.L. Parry *et al.* (eds), *The Impact of Climatic Variations on Agriculture, Vol. 2: Assessments in Semi-Arid Regions*, Dordrecht, D Reidel.

Iwasaki, K. 1985. 'Regional classification of climatic change and variability in Australia since 1940' in H. Toya *et al.* (eds), *Studies of Environmental Changes due to Human Activities in the Semi-Arid Regions of Australia*, Tokyo Metropolitan University.

Kraus, E.B. 1954. 'Secular changes in the rainfall regime of the south-east of Australia, *Q. J. R. Met. Soc.* 80, 591.

Lamb, P.J. 1985. 'Rainfall in sub-Saharan West Africa during 1941–83, *Zeitschrift für Gletscherkunde und Glazialgeologie* 21, 131–9.

Magari, K. 1980. 'Rainfall trend at Port Moresby from 1945 to 1976', *Weather* 35, 110–17.

Paltridge, G., Woodruff, S. 1981. 'Changes in global surface temperature from 1880 to 1977 derived from historical records of sea surface temperatures', *Mon. Wea. Rev.* 109, 2427–34.

Pittock, A.B. 1975. 'Climatic change and the patterns of variation in Australian rainfall', *Search* 6, 498–504.

Pittock, A.B. 1983. 'Recent climatic change in Australia: implications for a CO_2-warmed earth', *Climatic Change* 5, 321–40.

Russell, J.S. 1981. 'Geographic variation in seasonal rainfall in Australia — an analysis of the 80-year period 1895–1974', *J. Aust. Inst. Agric. Sci.* 47, 59–66.

Seelye, C.J. 1950. *Fluctuations and Secular Trend of New Zealand Rainfall*, Wellington, NZ

Meteorological Service.

Trenberth, K.E. 1977. *Climate and Climatic Change: a New Zealand Perspective* Wellington, NZ Meteorological Service.

Tucker, G.B. 1980. 'On assessing the climatic response to a continuing increase in the carbon dioxide content of the atmosphere' in G.I. Pearman (ed.) *Carbon Dioxide and Climate: Australian research*, Canberra, Australian Academy of Science.

Ward, W.T., Russell, J.S. 1980. 'Winds in south-east Queensland and rain in Australia and their possible long-term relationship with sunspot number', *Climatic Change* 3, 89–104.

Wright, P.B. 1974. 'Temporal variations in seasonal rainfall in southwestern Australia', *Mon. Wea. Rev.* 102, 233–43.

Chapter 26

Influence of severe local storms in south-east Queensland

C. Sisson and J.E. Hobbs   University of New England, Australia

Introduction

More severe storms occur in south-east Queensland than in any other part of Australia. In a recent report from the Insurance Council of Australia, listing major hazard events in Australia since June 1967, New South Wales recorded 13 severe storms, Victoria 25, and south-east Queensland 27. South-east Queensland covers an area of about 120,000 km^2 and has a population of about 1.5 million. The region experiences a subtropical climate with a summer maximum rainfall, although amounts show considerable spatial and temporal variation. In spring and summer (September–February) sporadic thunderstorms may account for up to 90 per cent of all rainfall in some districts.

The severe local storm hazard affects all areas within the region and causes millions of dollars of damage annually. The insurance pay-out figure for severe thunderstorm damage in south-east Queensland since June 1968 has been estimated at A\$327 million. This amount is an insurance assessment, so it is certainly a gross underestimation of the damage, as not all properties are insured, those that are may be inadequately covered, small claims are not included in the reported figures, and not all insurance claims are reported to the Insurance Council of Australia anyway. A true assessment of economic impact is therefore difficult to find.

The benefit of severe storms to the region is largely through the rainfall they provide, especially to the agricultural sector. However, such rainfall is often accompanied by strong winds and hail which wreak havoc on the environment. Brunt and Mackerras (1961) noted that in south-east Queensland both the heaviest short-period rainfall and the highest wind gusts are not found in association with tropical cyclones but tend to be of thunderstorm origin. Prentice *et al.* (1965) reported that hail was found in 7 per cent of all storms studied.

Thunderstorm incidence and characteristics

Data for 20 stations over a ten-year period, 1975–84, indicated a large spatial variation in the occurrence of thunderstorms. The Southern Border Highlands, between Toowoomba and Stanthorpe (Figure 26.1) had the highest incidence. Toowoomba had an annual mean of 22 events. Other pockets of relatively high incidence were Brisbane, Amberley, Sandy Cape and Monto. The stations recording fewest storms included Dalby with an annual mean of 6.2, Bundaberg with 8.1 and Southport with 11.3 (see Table 26.1).

Thunderstorm activity occurs predominantly in late spring and early summer with storms increasing markedly in October, peaking in November–December and

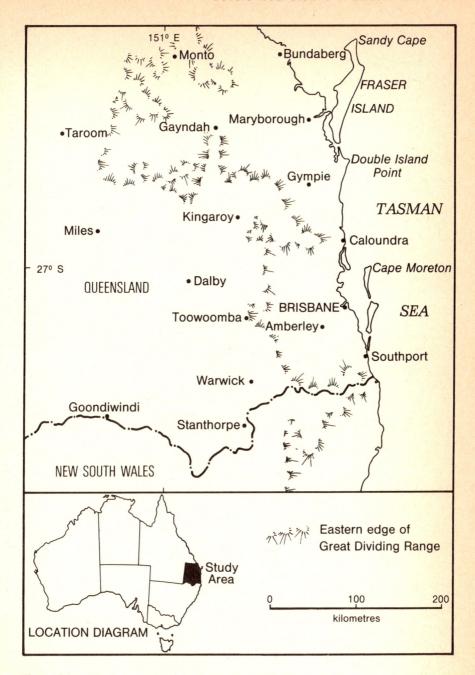

Figure 26.1   The study area of south-east Queensland

Table 26.1   Estimated frequency of thunderstorm events, 1975–84

Station	Jan.	Feb.	Mar.	Apr.	May	June	July	Aug.	Sept.	Oct.	Nov.	Dec.	Mean annual number
Amberley	22	13	21	10	2	2	–	2	11	22	41	45	19.1
Bundaberg	11	3	7	5	2	2	–	1	7	8	27	*	8.1
Caloundra	20	8	8	8	2	1	–	–	3	18	39	30	13.7
Cape Moreton	15	9	9	10	5	5	–	2	7	20	32	31	14.5
Dalby	5	6	4	1	1	2	–	2	4	11	13	13	6.2
Double Is. Pt.	20	14	15	17	7	5	1	1	7	20	43	35	18.5
Eagle Farm	18	17	13	8	4	6	1	1	10	30	50	50	20.8
Gayndah	25	20	14	15	2	3	–	3	13	27	44	*	18.4
Goondiwindi	23	21	11	4	2	–	2	2	9	18	24	27	14.3
Gympie	13	8	16	6	1	2	–	2	8	20	36	37	14.9
Kingaroy	19	11	17	8	3	2	–	1	12	18	28	48	16.7
Maryborough	25	18	14	9	4	3	1	2	6	19	37	33	17.1
Miles	16	19	12	7	2	2	1	6	9	24	41	40	19.9
Monto	35	22	16	11	4	3	1	1	7	21	49	30	21.0
Sandy Cape	24	15	21	23	11	5	9	5	16	20	59	46	25.4
Southport	13	13	8	7	1	–	–	–	6	15	26	24	11.3
Stanthorpe	34	21	18	10	7	2	–	1	7	25	47	48	22.0
Taroom	16	21	14	4	5	2	2	2	11	18	37	29	16.1
Toowoomba	30	28	26	14	3	4	2	2	15	32	62	65	28.3
Warwick	13	13	10	6	3	2	1	4	13	12	32	*	12.1

* Data missing for December, therefore annual means for these stations are probably under-estimates.
Source: Based on reports of thunderstorm activity plus analysis of rainfall records

dropping off through March to a minimum in July. Of the 20 stations studied over the ten-year period, six recorded storm peaks in December while the other 14 peaked in November. In general, the peak in the number of thunderstorms in November–December is slightly ahead of the temperature maximum, which occurs in January. The usual commencement of the wet season towards the end of that month, with its associated cloudiness, probably accounts for the earlier peak in thunderstorms.

Most thunderstorms occur between noon and midnight. Prentice *et al.* (1975) noted that 50 per cent occurred between 2 p.m. and 7 p.m. and 86 per cent between noon and midnight. An unpublished study of thunderstorms in the Brisbane area found that during the five-year period 1968–72, 53 per cent of all storms commenced during the six-hour period 3–9 p.m. while only 1 per cent commenced during the six-hour period 5–11 a.m. No time difference was noted in the commencement of frontal and non-frontal thunderstorms.

Thunderstorm movement through south-east Queensland is generally steered by winds at about 5000 m and seems to be rarely affected by winds below this level. Movement is usually along fairly straight paths, with the majority of storms moving from the south-west.

Thunderstorms, as a major rain-producing influence within south-east Queensland, have important implications for hydrology and agriculture. The contribution of thunderstorm rainfall to total rainfall amounts across the study area ranged from an annual average of 48.6 per cent at Gayndah to only 11 per cent at Dalby. Toowoomba registered the greatest mean annual thunderstorm rainfall

Table 26.2 Annual mean thunderstorm rainfall, 1975–84

Station	Mean annual number of storms	Annual mean rainfall (mm)	% due to thunderstorm activity
Amberley	19.1	878.3	31.0
Bundaberg	8.1	934.8	17.2
Caloundra	13.7	1388.8	15.1
Cape Moreton	14.5	1445.3	19.8
Dalby	6.2	730.1	11.0
Double Is. Pt	18.5	1489.6	26.5
Eagle Farm	20.8	1096.3	23.7
Gayndah	18.4	734.8	48.6
Goondiwindi	14.3	684.5	27.7
Gympie	14.9	973.9	34.7
Kingaroy	16.7	818.1	34.4
Maryborough	17.1	1061.7	22.3
Miles	17.9	1168.0	35.2
Monto	21.0	663.6	42.1
Sandy Cape	25.4	1238.7	26.8
Southport	11.3	1406.9	12.1
Stanthorpe	22.0	831.5	35.8
Taroom	16.1	737.0	28.7
Toowoomba	28.3	1076.0	38.7
Warwick	12.1	727.4	25.4

with 417 mm, Dalby the least with only 81 mm (see Table 26.2). For all stations the annual average amount of rainfall associated with thunderstorm activity was 254 mm.

Attitudes, impacts and responses

Severe thunderstorms often strike without warning and the impact on communities is considerable. It is informative, therefore, to consider the attitudes of residents in the region to the severe local storm hazard in their area. In a survey of 94 residents, 69 from urban areas and 25 from rural areas, reflecting the proportion of urbanisation in the region, 33 per cent rated the hazard as a serious problem, 34 per cent as a nuisance problem, 31 per cent believed the hazard caused little trouble, and only 2 per cent thought storms caused no trouble. None of the respondents replied they did not know. Rural dwellers apparently had a higher level of concern that those in urban areas. Forty-eight per cent of rural dwellers thought storms to be a serious problem compared with only 25 per cent of urban dwellers; 36 per cent of rural residents and 37 per cent of those in urban areas thought the hazard a nuisance problem; 16 per cent of rural dwellers and 35 per cent of the urban sector found storms caused little trouble. All of those feeling that severe storms caused no trouble came from the urban sector. Of the 94 respondents, 52 per cent had actually experienced loss or damage from severe local storms, 84 per cent of the rural dwellers and 40 per cent of the urban dwellers. Of the overall sample, 68 per cent expected severe local storms to cause future damage in their area.

Insurance pay-out figures give some indication of damage sustained from severe

Table 26.3   Estimated insurance losses for severe storms in south-east Queensland

Date		Event	Original dollars	1985 dollars
June	1967	Storms, Brisbane	approx. 5m	21.75m
Nov.	1968	Storm, Killarney	More than 1m	4m
Dec.	1972	Storms, Brisbane	More than 1m	3.25m
Nov.	1973	Tornado/storms, Brisbane, SW Qld	200m damage	500m+
Jan.	1976	Hailstorm Toowoomba	12m pay-out	30m
Feb.	1976	Storm Bundaberg	3m pay-out	6.75m
Dec.	1977	Storms Brisbane	1.75m	4–5m
Dec.	1980	Storm, Brighton	15m	60m
Feb.	1981	Floods/storms, Dalby	10m	30m
Nov.	1981	Storms, Beaudesert & Nambour	1m	1.25m
Nov.	1981	Storms, North Sunshine Coast	1m	1.25m
Jan.	1983	Hail & wind storm, Chinchilla	2m	2m
Sept.	1983	Windstorm, Brisbane	100,000	250,000
Dec.	1983	Storm, Ipswich	500,000	1m +
Apr.	1984	Storm, Brisbane	2m	2m
Dec.	1984	Storms, Roma	2.5m	2.5m
Jan.	1985	Storms, Brisbane	150m	175m +
Feb.	1985	Storms, Gympie	1.5m	1.5m

Source: Insurance Council of Australia

local storms (see Table 26.3). The storm that hit Brisbane on 18 January 1985 lashed the centre of the city with winds of 120–160 km h^{-1}, blinding rain and the largest hailstones ever recorded in Queensland. The storm lasted for about 20 minutes, turning daylight into darkness and reducing visibility to zero. It struck at 4.50 p.m. on a Friday afternoon as many workers were returning home from work, causing traffic to come to a complete standstill and many cars to receive severe hail damage. The total number of insurance claims amounted to 105,355, with a pay-out figure of over A$177 million (see Table 26.4). New and used car yards often fare badly in severe storms. One major distributor in the Brisbane suburb of Eagle Farm reported the damage listed in Table 26.5 from the 18 January storm.

Rural south-east Queensland is hit equally hard by the severe storm hazard. A series of storms, caused primarily by the passage of an upper pool of cold air, stretched from the state border in the south to Kingaroy in the north on Christmas Day 1986. Up to 175 mm of rain was received in some areas and hail, ranging in size from golf balls to cricket balls, was reported. The hail and wind shredded crops, devastated grazing land, and in some areas, stubble was mulched to ground level. The Queensland Graingrowers Organisation received over 100 claims for damage up to A$250,000. However, it was reported that only 30–40 per cent of growers were insured and approximately 16,000–20,000 ha of maize, mung beans, sunflower, sorghum and canary seed worth A$5 million were damaged. Most crops were at the flowering stage and had some chance of recovery although yield would be reduced. On the Granite Belt, in the Stanthorpe area, damage was estimated at A$3 million with at least 30 per cent of fruit growers incurring damage to over 50 per cent of their crops. Stanthorpe can have up to 60 storms per year across the district, with crop losses as high as 50 per cent, although 20–30 per cent losses are more common. Hail damaged apples can be juiced, but there is no outlet for stoned fruit and vegetables.

The Barley and Wheat Marketing Boards operate compulsory hail insurance

Table 26.4 Storm damage, Brisbane city and suburbs, 18 January 1985

Category	No. of claims	Estimated losses paid and outstanding as at 30 September 1985 (A$)
Commercial/industrial	6,027	47,919,541
Domestic	69,578	87,618,698
Motor vehicle	29,512	41,285,061
Other	238	514,256
Total	105,355	177,337,556

Table 26.5 Storm frequency to stock of a Brisbane car dealer, 18 January 1985 (AS$)

Damage to 34 registered company vehicles	46,858
Damage to 421 stock vehicles	158,571
Damage to buildings	99,909
Damge to 145 vehicles at wharf	396,237
Damage to trading	25,690
Total	727,265

Table 26.6 Examples of crop losses from severe storm damage

1. *Barley Marketing Board*
   1986–7 barley season:
       9552 ha affected = loss of 5,227 tonnes, compensation to growers A$417,296
   1985–6 barley season:
       35,000 tonnes lost, compensation of A$2,400,000

2. *State Wheat Board*
   Damage to Authority installations, temporary storage sites, plant and equipment in south-east Queensland since 1981 = A$499,000
   Loss in tonnes since 1981 = 151,847 (annual mean = 21,692)
   Compensation to growers = $14,371 (annual mean = A$2,053)

schemes to compensate for hailstone damage to crops (as opposed to other storm damage). Growers are required to register their crops before a certain date each year (August for barley, September for wheat) and have to carry an excess themselves (for barley the first 60 kg ha^{-1} of loss involved, for wheat the first 75 kg ha^{-1}). The scheme is operated on a pool basis, levies and rates being determined by the Boards each year, dependent upon the money available in the compensation fund, the quality of the crop harvested and the quantity of the crop damaged by hail (see Table 26.6).

Severe thunderstorms in south-east Queensland also disrupt essential services within the region. The South-East Queensland Electricity Board rates the physical impact of storms by the number of consumers blacked out by each event (see Table 26.7). During the eight years from 1 July 1977 to 22 August 1985, there were 102 storms in Brisbane that caused damage to the electricity distribution system. The Queensland Electricity Commission, which supplies in bulk to the different electricity boards, also reports damage due to severe storm activity. Their power lines

Table 26.7 Blackouts caused by severe local storms

Major storms: Date	Approx. no. of connections blacked out in Brisbane	Remarks
16.12.77	40,000	Electrical storm
7.11.78	10,000	Electrical storm
13.12.78	20,000	Electrical storm
27.4.79	10,000	Electrical storm
20.11.79	40,000	Electrical storm
20.1.80	35,000	Electrical storm
22.11.80	12,000	Electrical storm
16.12.80	20,000	Electrical storm
20.12.81	15,000	Electrical storm
6.12.82	14,000	Electrical storm
7.12.82	18,000	Electrical storm
22.6.83	16,000	High winds
18.4.84	30,000	Mini cyclone*
3.7.84	12,000	High west winds
5.11.84	10,000	Electrical storm
6.1.85	25,000	Electrical storm
18.1.85	80,000	Severe hail storm
2.9.85	10,000	High winds
17.10.85	Electrical storm	
30.11.85	10,000	Electrical storm
9.3.85	10,000	Electrical storm

* Probably a tornado
Source: South-east Queensland Electricity Board

have overhead earth wire shielding to protect them from serious over-voltages caused by lightning proximity strikes and have a high clearance from the ground to minimise damage from wind-blown debris during storms, yet from 1 July 1977 to 5 May 1987 there were 183 breakdowns on the Commission's high-voltage transmission lines in south-east Queensland directly attributed to storm activity.

It is interesting to consider the attitudes of local residents concerning who should be responsible for financial aid in the case of severe loss or damage due to storms. The federal government in Australia provides A\$3 in aid for every A\$1 provided by the states above a base amount. However, this grant does not usually apply to private assets and dwellings. In the survey conducted, 47 per cent of the respondents believed in government responsibility (federal, state and local), 35 per cent though assistance should be provided through insurance and 18 per cent felt the individual should be responsible.

Conclusions

The severe local storm hazard has widespread and variable impacts in south-east Queensland. Large financial losses and major structural damage are common. The major problems arising from severe local storms still concern their relative unpredictability, coupled with the fact that single storm events tend to cause damage over relatively small areas. Preparedness is in most circumstances impossible within an affordable framework. The fact is that in most instances enterprises and communities have to learn to live with such hazards and be prepared to stand any losses incurred, subject to reduction by whatever insurance may be available. Even given adequate warning there is little that can be done to protect entire crops or communities. At the level of the individual household or enterprise, some protection is possible but not always affordable. Perhaps the best that can be hoped for is an improvement in the awareness of the nature of the hazard and of the likelihood that it may strike just about anywhere at any time. With current attitudes towards natural disasters in Australia there is always the likelihood that government assistance will in fact help to meet losses incurred, but it is arguable whether this is a desirable approach.

References

Brunt, A.T., Mackerras, D. 1961. 'A study of thunderstorms in south east Queensland', *Aust. Met. Mag.* 34, 15–28.

Prentice, S.A. *et al.* 1965. *A Ten Year Survey of Thunderstorms in Queensland*, Department of Electrical Engineering, University of Queensland.

Chapter 27

The greenhouse effect and future climatic change*

A. Barrie Pittock   CSIRO, Mordialloc, Australia

The greenhouse effect

The 'greenhouse effect' is the name given to a global warming expected to be brought about by an increase in the atmospheric concentration of carbon dioxide CO_2 and other gases which allow sunlight to reach the earth's surface but prevent some of the infrared or heat radiation given off by the earth from escaping into space.

This possibility was first suggested in 1896, but it is only since 1958 that accurate measurements of the increasing concentrations of CO_2 in the atmosphere have been made. These show a fairly regular annual rate of increase of about 4 per cent which is primarily due to the burning of fossil-fuels (oil, coal and natural gas). Measurements of past concentrations in bubbles of air trapped in ice cores from Antarctica to Greenland show that pre-industrial concentrations were about 270 parts per million by volume (ppmv). The present concentration is about 348 ppmv, and a doubling of the pre-industrial level is expected by the latter half of the twenty-first century.

There is some uncertainty as to the probable effect of a given increase in CO_2 concentration on the climate, due to the great complexity of the climate system. A doubling of CO_2 would, on theoretical grounds, lead to an average surface warming of about 2–4 °C, with greater warmings occurring in winter and at high latitudes due to the reinforcing effect (positive feedback) of warmer temperatures leading to less snow and ice cover and thus to more sunlight being absorbed at the surface (DOE 1985a; Bolin *et al.* 1986).

This warming effect is due to the relative opacity of CO_2 to infrared or heat radiation, so that energy absorbed from sunlight has greater difficulty escaping into space as heat radiation. A major uncertainty arises from the possibility of either positive (reinforcing) or negative (damping) feedbacks due to changes in cloudiness. The effect of changing cloud cover depends on the height, latitude and season of the cloudiness.

New factors to be taken into account

Several factors have led to greater interest in and concern about the greenhouse effect in the last few years. These factors include:

- a growing consensus amongst atmospheric scientists that increased infrared

---

*This contribution is an adaptation of a paper that first appeared in *Engineers Australia*, 6 February 1987, 40–3.

absorption by the atmosphere will lead to appreciable surface warming, despite cloud feedback effects;

- the realisation that increases in infrared absorbing gases other than CO_2 will cause an equivalent doubling of CO_2 on a much shorter time-scale than CO_2 alone (Ramathan *et al*. 1985);
- new evidence from ice cores that pre-industrial levels of CO_2 and some other infrared-absorbing gases, notably methane, were much lower than present concentrations (Pearman *et al*. 1986);
- a growing realisation of the many ways in which a greenhouse warming might impact on human society;
- observations of increases in global mean surface air and sea surface temperatures, and of rises in mean sea level, which are consistent with those to be expected from the increase in concentrations of greenhouse gases to date (Bolin *et al*. 1986).

The Villach Statement

In 1985 the Scientific Committee on Problems of the Environment, which is a standing committee of the International Council of Scientific Unions, produced a major report entitled *The Greenhouse Effect, Climatic Change, and Ecosystems*. A draft of this report was discussed at an international meeting of scientists at Villach in Austria in October 1985. This resulted in what has become known as the 'Villach Statement', which is reproduced in the final version of the report (Bolin *et al*. 1986). The Villach statement opens by saying that:

> As a result of the increasing concentration of greenhouse gases, it is now believed that in the first half of the next century a rise of global mean temperature could occur which is greater than any in man's history.

The statement goes on to say:

> Many important economic and social decisions are being made today on long-term projects — major water resource management activities such as irrigation and hydro-power, drought relief, agricultural land use, structural designs and coastal engineering projects, and energy planning — based on the assumption that past climatic data, without modification, are a reliable guide to the future. This is no longer a good assumption . . . It is a matter of urgency to refine estimates of future climate conditions to improve these decisions.

The scientists at the Villach meeting agreed that, if present trends continue, the combined effect of all the greenhouse gases would be equivalent to a doubling of CO_2 concentration as early as the 2030s, with a consequent equilibrium surface warming of between 1.5° and 4.5 °C. This warming could be delayed by one or two decades due to the large heat capacity of the upper layers of the oceans. Such an equilibrium warming is estimated to lead to a rise in global sea level of 20–140 cm, due essentially to thermal expansion of the oceans rather than to the melting of the ice sheets, which would take many hundreds of years.

Regional assessments

Before an attempt is made to outline what physical, biological and socio-economic effects are likely at a regional level, some of the many uncertainties should be stressed. As the Villach Statement makes clear, even the global average picture has large uncertainties and error bars attached. When we come to look at detailed regional effects these become even greater.

Given a reasonable scenario for the growth in concentration of greenhouse gases due to the burning of fossil-fuels and other agricultural and industrial processes, we are forced to rely on the results of computer models of the global climate incorporating increased infrared absorption in the model atmosphere. There are many such models which have been applied to the problem, but they all have simplifications due to our lack of detailed knowledge of some parts of the climate system, and the expense of running more complex models. These models provide what are believed to be reasonably reliable predictions of the gross climate change to be expected, but they are not reliable when it comes to geographical detail relating to a particular region.

The computer model studies are therefore the starting point, but for regional impact analysis we must supplement these with various other lines of evidence such as analogies with past warmer intervals in the instrumental or palaeoclimatic record, and various lines of reasoning from our knowledge of atmospheric and oceanic properties.

The result for any given region is that some features of a greenhouse-warmed climate can be foreseen reasonably reliably, but other features are less certain and may vary from locality to locality in ways we do not yet fully understand. Here I will illustrate this with a climate scenario for Australia. It is meant for guidance as to possible regional effects, especially as to what factors may be important, but the details as they apply to a particular locality (even within Australia) are as yet rather uncertain.

Physical effects: An Australian scenario

Broad physical effects are likely to be felt in Australia by the year 2030. In inland areas the temperature increases are expected to be about 2 °C in northern Australia and up to 3° or 4 °C further south. Near the coast these increases could be moderated by the lagging behind of the sea-surface temperatures, and it is probable that winter and overnight minimum temperatures will warm more than summer and daily maximum temperatures.

Rainfall changes will vary with season and location. Based on observed changes so far this century, analysis of palaeoclimatic data, and other evidence (Pittock 1983), it would seem likely that in general the summer rainfall regime will intensify and push further south. This has already happened through central NSW from the Hunter valley west through Dubbo, where average spring, summer and autumn rainfall in some months has increased by 30–40 per cent since early this century (see Figure 27.1a).

On the other hand, winter rain associated with the fronts embedded in the westerlies at middle latitudes is expected to occur further south, leading to reduced rainfall especially in the south-west of Western Australia. Observed decreases in this region so far this century are about 10–20 per cent (see Figure 27.1b). Such

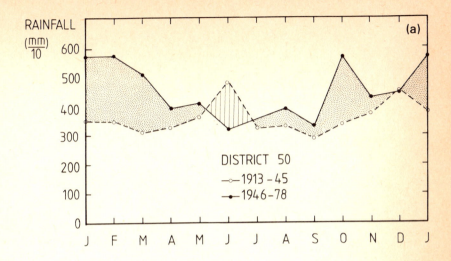

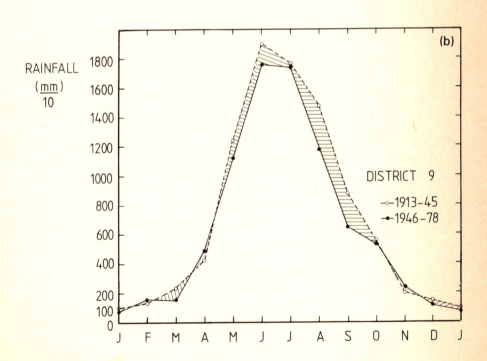

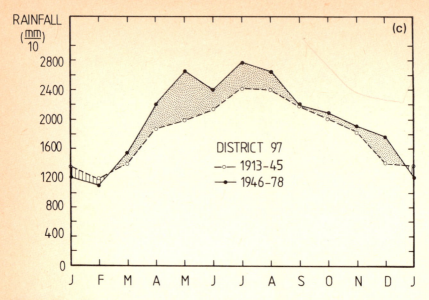

Figure 27.1   Mean annual cycles of rainfall in Australia in the intervals 1913–45 and 1946–78 for districts 50, 9 and 97
(a) district 50 around Dubbo in central New South Wales
(b) district 9 around Perth in Western Australia
(c) district 97 in western Tasmania

Note: Increases between the two intervals are indicated by stippling, and decreases by hatching (from Pittock 1983).

trends are expected to continue as the greenhouse effect intensifies.

Rainfall in western Tasmania, however, has increased in recent decades (Figure 27.1c). Possibly this is due to a transient effect associated with the continent warming up faster than the Southern Ocean, thus leading to a strengthening of the westerlies south of the mainland, but this is very uncertain.

The implications for rainfall in south-eastern Australia are not clear at present. It is clear, though, that increased surface temperatures make it possible for the air to hold a greater absolute amount of water vapour, so that in situations where this air is continuously lofted to create intense rainfall, greater maximum rainfall rates are possible. Thus the maximum possible daily rainfall will increase. This is a critical factor in the intensity and frequency of flash floods, and in the design of dam spillways intended to cope with maximum run-off in water catchments.

Soil moisture, and water run-off available for urban and rural water supplies, river flow and flooding, are due to an often delicate balance between rainfall and evaporation. In general wind speeds might be expected to be slightly reduced, due to reduced north–south differences in temperature, but higher temperatures may still lead to a general increase in evaporation. With marked regional and seasonal changes in rainfall, such changes will lead to marked regional and seasonal changes in the water balance (see, for example, Gleick 1987).

Tropical cyclones (or hurricanes, or typhoons) are born and sustained only where the sea-surface temperature exceeds about 27 °C, though, once generated, they can degenerate into tropical rain depressions which can move inland and much further polewards. An increase in sea-surface temperature of 2–3 °C would enable tropical cyclones to form 200–400 km further south. It would thus be perfectly possible for a cyclone to strike Brisbane in the year 2030. Whether or not tropical cyclones might also increase in intensity and frequency has yet to be established (but see Miller 1958; and Emanuel 1987). Changes in ocean circulation could, however, substantially alter this picture, and indeed other aspects of regional climate.

Short-lived climatic extremes such as floods and droughts normally occur as part of statistical fluctuations about some average value. In general, a shift in the average value towards one extreme would lead to a disproportionate increase in the frequency of occurrence of what were classed as extremes in that direction, and to a large decrease in the frequency of extremes at the other end of the distribution. Thus the recurrence interval for various extreme events might change markedly for relatively small changes in the mean.

Changes in the water balance will also lead to changes in the water table. Where rising water tables are indicated, this could lead to increasing problems of soil salinity, particularly in inland river basins.

Higher temperatures will in general raise the equilibrium snowline by about 100 m °C^{-1}. However, there are complications when considering alpine snow areas in Australia, since these are for the most part not in the areas above the equilibrium snowline. In the snow season the presence of snow is due to a dynamic balance between winter snowfall and progressive melting or washing away of snow when temperatures reach above freezing between snowfalls. So if the frequency of cold outbreaks, sufficiently cold to allow snow to fall, were actually to increase (which is possible in south-eastern Australia), higher average temperatures may not initially reduce the winter snow-covered area. It would, however, shorten the snow season, and if the warming became great enough, snow could be eliminated.

Sea-level rise is another major consideration. The Villach Statement suggests a rise of 80 > 60 cm due to an equilibrium global temperature increase of 1.5–5.5 °C, which is expected a few decades beyond 2030 (allowing for the thermal lag in the oceans). This is based on the observed average rise in sea level over the past 100 years of about 10 cm, which closely parallels a global average rise in surface temperature of about 0.5 °C (see Figure 27.2). It is believed that this is almost entirely due to thermal expansion of the oceans (Robin, in Bolin *et al*. 1986).

There are great uncertainties, however, in these estimates, with some experts suggesting smaller rises (see, for example, Revelle 1983). Local changes in sea level relative to the land can occur for a variety of other reasons, such as consolidation of silt, withdrawal of groundwater, and actual tectonic uplift or sinking of land masses. For instance, at Port Adelaide the increase in sea level relative to land is thought to average about 1.5–2 mm per year.

The Villach Statement is suggesting up to ten times this rate of rise in future. Clearly, if this were to happen there would be a rapid loss of beaches, some coastal flooding, erosion and increasing storm damage (Titus 1986a). There could also be problems with drainage of low-lying coastal areas, and with increased salinity of ground water.

Finally, increased ambient carbon-dioxide concentration would directly affect plant growth. Laboratory and glasshouse experiments on individual plants indicate that a doubling of CO_2 would lead to a 0–10 per cent increase in growth and yield for C_4 plants such as corn, sorghum and sugarcane, and a 10–50 per cent increase

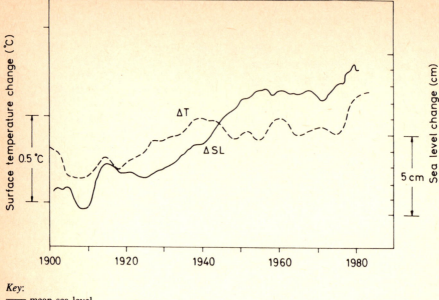

Key:
—— mean sea level
------ mean surface temperature

Figure 27.2   Graphs of globally averaged observed changes in mean sea level and mean surface temperature since 1900

Source: Bolin *et al.* (1986).

for C_3 plants which include wheat, rice, barley and many fruits and vegetables (DOE 1985b).

It is not clear that this increase will occur, however, in whole crops or forests due to canopy and other effects. Other complications include the fact that various weeds will also benefit, and that increased growth will not necessarily go into the edible part of crops or increase the amount of protein produced. Nevertheless, the increased water-use efficiency of plants growing under high CO_2 conditions will probably extend their range into more arid areas.

Biological effects

Based on the foregoing scenario, a number of biological and socio-economic effects may be postulated (Titus 1986b). Most are 'common-sense' observations, not presently backed up by much scientific research. It is important that these be recognised early, so that scientific research can be undertaken to confirm, amend and quantify them.

As stated above, carbon dioxide is well known to have a 'fertiliser' effect, stimulating plant growth. The higher ambient CO_2 concentrations which favour some species over others, together with climatic changes, will lead to changes in species composition in natural vegetation, and in many farming situations. It may be necessary in some areas to switch from one crop to another and to change weed-control strategies.

One might also expect that there will be climate-related changes in the occurrence of various plant diseases, such as rust in wheat and various fungal diseases, and in insect pests such as locusts and aphids.

Rising water tables in some areas may lead to increased salination problems and consequent loss of farmland. This may happen both in various inland river basins and in low-lying coastal areas due to sea-level rise. Coastal zone flooding and erosion will also particularly affect mangroves and river estuaries, which are important spawning grounds for fish.

Particularly in limited areas where sharp climatic gradients occur, such as is the case in many nature reserves designed to protect endangered species, changing climate may render the reserves inappropriate and threaten the extinction of some species of plants and animals.

Socio-economic effects

The changing climate is likely to have a number of socio-economic effects, the severity of which will increase with the rate of change, since economic costs are most serious when changes occur more rapidly than the design life of the systems concerned. An example would be the spillway design for a large dam. It should be emphasised that the following comments are very preliminary — improved estimates will need to be made by engineers and scientists in the relevant fields.

The climatic and biological effects on crops will have important consequences for many rural communities. Some will be better off, others will be able to adapt by changing their farming strategies, with different scheduling of planting and harvests, changing varieties of wheat, or switching to some other grain crop or from beef to sheep or vice versa.

However, some marginal areas will become uneconomic, and other areas will remain profitable only with large capital expenditure on equipment for irrigation or flood control and drainage, or changes in farm machinery and infrastructure. Significant population shifts could result from these changing circumstances.

Coastal management, depending on the magnitude of the sea-level rise, may be greatly affected. Remedial action could be extremely costly, with major capital works necessary to protect low-lying areas, and especially water-front buildings, marinas, ports, and industries sited near coastlines for cooling water or ease of freight handling. Beaches and holiday resorts may be badly affected.

Urban and rural water supply authorities will face major problems in allocating water, ensuring the safety of dams subject to greater peak river flow, and in planning for adequate water supply under progressively changing conditions. Related to this is the management of hydroelectric generation, which may be particularly affected by reductions in snow storage through the spring and summer months, and changes in seasonal demand patterns. Ski resorts would almost certainly be adversely affected.

Conservation of animal and plant varieties will be a serious issue, as some nature reserves and national parks become increasingly inappropriate climatically for the species they were meant to protect. Loss of generic diversity which might follow could have long-term implications for plant breeding and medical research and development.

Domestic energy demand could also change significantly, with less demand for winter heating and more for summer cooling.

In Australia and elsewhere coal, oil and chemical industries could also be seriously affected by any attempts to regulate the emissions of greenhouse gases, either domestically or overseas, and by changing economic circumstances and product demand. It is likely that other local industries will also experience significant economic impacts, some favourable and others less so, due to changing competition and demand overseas. For example, it is presently widely believed that grain production in North America and Europe could be drastically affected by reductions in summer soil moisture. Initially this would be expected to increase the demand and prices for Australian grains, but, if the economic effects overseas lead to a major economic crisis, this may not follow in the longer term.

Finally Australia, along with other countries, will be faced with the question of what stand to take on the question of international regulation of fossil-fuel usage and the production of other greenhouse gases.

The recently concluded Montreal Protocol under the International Convention for the Protection of the Ozone Layer provides a hopeful precedent in this regard (Johnston 1987). Nevertheless, international disagreements may well arise as to the most appropriate policies to be pursued, and the enforcement of any such policies.

In view of the probable need to reduce total fossil-fuel usage in an attempt to slow down the greenhouse effect, the issues familiar from the oil crisis — energy conservation, alternative (non-fossil-fuel) energy sources and renewable energy resources — assume a new and more lasting significance.

Conclusion

It is highly probable that over the next several decades climatic changes of an unprecedented magnitude in human history will take place. These will have numerous and important consequences at a regional level. We can either wait to see what happens and react as best we can, or we can try to anticipate the effects and plan the best strategy to maximise the gains and minimise the losses. The latter strategy will require the energetic collaboration of climatologists with geographers, engineers, agricultural scientists, and many others. Where possible we should seek to slow down the changes and avoid the worst possible outcomes. In view of the present uncertainties, this will require intensive research, periodic reassessment of the situation, and the development of resilient and adaptable development strategies.

References

Bolin, B., Doos, B.R., Jager, J., Warrick, R.A. (eds) 1986. *The Greenhouse Effect, Climatic Change, and Ecosystems*, Chichester, John Wiley.

DOE 1985a. *Projecting the Climatic Effects of Increasing Carbon Dioxide*. Washington, DC, US Department of Energy, DOE/ER-0237.

DOE 1985b. *Direct Effects of Increasing Carbon Dioxide on Vegetation*. Washington, DC, US Department of Energy, DOE/ER-0238.

Emanuel, K.A. 1987. 'The dependence of hurricane intensity on climate', *Nature* 326, 483–5.

Gleick, P.H. 1987. 'Regional hydrologic consequences of increases in atmospheric CO_2 and other trace gases', *Climatic Changes* 10, 137–61.

Johnston, K. 1987. 'Looking ahead to the greenhouse after ozone agreement reached', *Nature* 329, 277.

Miller, B.I. 1958. 'On the maximum intensity of hurricanes', *J. Meteorol.* 15, 184–95.

Pearman, G.I., Etheridge, D., De Silva, F., Fraser, P.J. 1986. 'Evidence of changing concentrations of atmospheric CO_2, N_2O and CH_4 from air bubbles in Antarctic ice', *Nature* 320, 248–50.

Pittock, A.B. 1983. 'Recent climatic change in Australia: Implications for a CO_2-warmed earth', *Climatic Change* 5, 321–40.

Ramathan, V., Cicerone, R.J., Singh, H.B., Kiehl, J.T. 1985. 'Trace gas trends and their potential role in climatic change', *J. Geophys. Res.* 90, D3, 5547–66.

Revelle, R.R. 1983. 'Probable future changes in sea level resulting from increased atmospheric carbon dioxide' in *Changing Climate*, Report of the Carbon Dioxide Assessment Committee, Washington, DC, National Academy Press 433–48.

Titus, J.G. 1986a. 'Greenhouse effect, sea level rise, and coastal zone management', *Coastal Zone Management Journal* 14, 147–71.

Titus, J.G. (ed.) 1986b. *Effects of Changes in Stratospheric Ozone and Global Climate. Vol. 3: Climatic Change*, Washington, DC, United Nations Environment Programme and US Environmental Protection Agency.

Appendix

Addresses of contributors who were participants

Professor W. Bach
Centre for Applied Climatology and Environmental Studies
Department of Geography
University of Münster
Robert-Koch Str. 26
4400 Münster
FEDERAL REPUBLIC OF GERMANY

Dr R. Brázdil
Katedra Geografie
University J. E. Purkyně
Kotlářská 2
611 37 Brno
CZECHOSLOVAKIA

Professor C.N. Caviedes
Department of Geography
University of Florida
Gainesville, FL
USA

Dr T.D. Davies
Climate Research Unit
University of East Anglia
Norwich
NR4 7TJ
UNITED KINGDOM

Professor A. Douguedroit
Institut de Géographie
Université d'Aix-Marseille II
29 Avenue Robert Schuman
13621 Aix-en-Provence
FRANCE

Dr L.M. Druyan
National Aeronautics and Space Administration
Goddard Space Flight Center
Goddard Institute for Space Studies
2880 Broadway
New York
NY 10025
USA

Dr G. Farmer
Climatic Research Unit
University of East Anglia
Norwich
NR4 7TJ
UNITED KINGDOM

Professor Y. Goldreich
Department of Geography
Bar-Ilan University
52100 Ramat-Gan
ISRAEL

Professor S. Gregory
Department of Geography
University of Sheffield
Sheffield
S10 2TN
UNITED KINGDOM

Dr. J.E. Hobbs
Department of Geography and Planning
University of New England
Armidale NSW
AUSTRALIA

Dr M. Hulme
Department of Geography
University of Salford
M5 4WT
UNITED KINGDOM

Dr J. Jacobeit
Lehrstuhl für Physische Geographie
University of Augsburg
Universitätstr. 10
8900 Augsburg
FEDERAL REPUBLIC OF GERMANY

Dr P.D. Jones
Climatic Research Unit
University of East Anglia
Norwich
NR4 7TJ
UNITED KINGDOM

Dr J.A. Owen
Meteorological Office (MO 13)
London Road
Bracknell
Berkshire
RG12 2SZ
UNITED KINGDOM

Mr D.E. Parker
Meteorological Office (MO 13)
London Road
Bracknell
Berkshire
RG12 2SZ
UNITED KINGDOM

Dr A.H. Perry
Department of Geography
University College of Swansea
Singleton Park
Swansea
SA2 8PP
UNITED KINGDOM

Dr A.B. Pittock
CSIRO Division of Atmospheric Physics
Private Bag 1
Mordialloc
Victoria 3195
AUSTRALIA

Professor Dr C.-D. Schönwiese
Institute for Meteorology and Geophysics
Goethe-University
Feldbergstr. 47
6000 Frankfurt 1
FEDERAL REPUBLIC OF GERMANY

Professor D. Sharon
Institute of Earth Sciences
Hebrew University of Jerusalem
91904 Jerusalem
ISRAEL

Professor P.D. Tyson
Climatology Research Group
The University of the Witwatersrand
1 Jan Smuts Avenue
2001 Johannesburg
SOUTH AFRICA

Dr M.N. Ward
Meteorological Office (MO 13)
London Road
Bracknell
Berkshire
RG12 2SZ
UNITED KINGDOM

Professor M.M. Yoshino
Institute of Geosciences
The University of Tsukuba
Ibaraki 305
JAPAN

Index